CHANGE AND CONTINUITY IN AGRARIAN RELATIONS

About the Author

Gopal Krishna Karanth (*b.*, 1952) obtained Masters in Sociology from Bangalore University and doctoral degree from Jawaharlal Nehru University, New Delhi. For his Ph.D., he carried out an intensive village study under the supervision of Professors M.N. Srinivas and T.K. Oommen. He is currently an Associate Professor of Sociology at the Institute for Social and Economic Change, Bangalore. His interests include agrarian sociology, sociology of development and research methodology.

He is the author of *Rural Youth : A Sociological Study of a Karnataka Village* (Concept 1981), and several research papers in important journals. Currently, he is engaged in the study of Entrepreneurship and Social Mobility among Scheduled Castes; and Rural Livelihood Systems and Sustainable Resource Management.

Change and Continuity in Agrarian Relations

G.K. KARANTH
Institute for Social and Economic Change
Bangalore

Foreword by

JAMES MANOR
Director
Institute of Commonwealth Studies
University of London
London

CONCEPT PUBLISHING COMPANY, NEW DELHI - 110059

ISBN 81-7022-555-8

First Published 1995

Published and Printed by

Ashok Kumar Mittal
Concept Publishing Company
A/15-16, Commercial Block, Mohan Garden
New Delhi-110 059 (India)

DEDICATED
WITH LOVE AND AFFECTION
TO
AIRODY SEETHARAMA KARANTH
AND
SANJIVI KARANTH

Foreword

A few years ago, the distinguished sociologist M.N. Srinivas told me with great delight that he and Professor T.K. Oommen of the Jawaharlal Nehru University, New Delhi, were supervising a young anthropology student who showed great promise in his early fieldwork in a Karnataka village. The young man appeared to have the tact and good humour to develop close ties to a diversity of villagers and to extract invaluable information from them, and the intelligence and sensitivity to make perceptive use of it.

He was referring to the author of this book, and the work that G.K. Karanth has produced—here and elsewhere—clearly indicates that the early promise which Srinivas had recognised and cultivated in his gentle way has been fulfilled. I was lucky enough to observe Karanth interacting with residents of the village which is assessed here, and the empathy and insight which he brought to his work was vividly apparent in a very short time.

This volume offers us a highly revealing portrait of the social world of an Indian village, as it changes over time. It does not represent a return to the kind of study which understandably pre-occupied the first wave of anthropologists and sociologists who undertook work in rural India, although it shares many of their virtues. The focus here is a village interacting with the wider world. The Influence of competitive politics, of attempts by the state to promote land reform and the welfare of disadvantaged groups, and of technological change are much in evidence. So is the complex, ambiguous nature of the resulting changes.

Through it all, the author has retained a sympathetic under-standing of the people of this village, and of the resilience of pre-existing social order there. He does not flinch from reminding us of persisting inequalities and injustices, but the interdependence of social groups there and occasional acts of humanity are also evident.

The result is a study that is anchored in the rich tradition of Indian anthropology and sociology, which will also be of interest to historians, economists and political scientists, on whose work the author wisely draws.

JAMES MANOR
Director
Institute of Commonwealth Studies
University of London

Acknowledgements

This book is based on my doctoral dissertation, which I prepared under the guidance of Professor T.K. Oommen at the Centre for the Study of Social Systems, Jawaharlal Nehru University, New Delhi. Professor Oommen has been a constant source of inspiration which enabled me to analyse Rajapura findings in the change and continuity perspective. I remain indebted to him for guiding me through the doctoral studies and for the moral support I have received in all my endeavour. Mrs. Oommen was very kind and hospitable whenever I visited them at their house for dicussions with Professor Oommen.

Fieldwork for the study was carried out when I was a student at the Institute for Social and Economic Change, Bangalore. I was particularly fortunate in having Professor M.N. Srinivas to supervise my fieldwork. It was Professor Srinivas who made me realise the importance of fieldwork and its relation to theory. I am grateful to Professor Srinivas and Mrs. Rukmini Srinivas for all the affection and kindness shown to me.

Mr. V.S. Parthasarathy inducted me into rural studies, spent several days with me in the field guiding and advising me in my work. He very patiently went through the earlier versions of this work and made suggestions to improve it. I am grateful to him. Professors A.M. Shah, B.S. Baviskar, G.S. Aurora, B.R. Chauhan, M.V. Nadkarni, Abdul Aziz gave me several comments and suggestions while revising this work. My friends George Mathew, H.G. Hanumappa and R. Sivaprasad helped me in finalizing the manuscript. My sincere thanks are due also to the faculty and friends at the Centre for the Study of Social Systems : Professors R.K. Jain, K.L. Sharma, Yogendra Singh, M.N. Panini, Nanduram, A.S. Hebbar, B.N. Rao and P.K. Jayswal.

At the Institute for Social and Economic Change, Mr. T.R. Satish Chandran and Dr. P.V. Shenoi provided me opportunities to frequently visit Rajapura to keep myself abreast of the latest there.

Professors Don Attwood, B.S. Baviskar, Jim Manor, Simon Charsley, Nick Hopkins, and Richard Wigg visited me at Rajapura during different stages of my fieldwork there. Discussions with them gave me many new ideas with which to look at the field-realities. I am indebted to them.

The people of Rajapura taught me a good deal more than what is reported in the book. To their warmth and affectionate hospitality, I remain grateful. I fondly recollect Mr. Govindaiah who introduced me to Rajapura when I was looking out for a village. Mr. B.G. Kulkarni prepared the map of Rajapura at a very short notice. Mr. Krishnachandra came to my rescue whenever the computer was not being cooperative or failed to understand my commands. Mr. T. Srinivas Murthy, Mr. T. Amarnath and Ms. Achala N.T. word-processed the manuscript with all care. The Indian Council of Social Science Research gave me a fellowships at ISEC and JNU to pursue the research. Dr. David Selvaraj of HEKS, Bangalore supported the publication of this book. Mr. Ashok Kumar Mittal and his colleagues at the Concept Publications were quite enthusiastic in undertaking the publication of this work. Professor James Manor has been a source of inspiration with his studies on Karnataka's society and polity. He readily consented to write a Foreword, when I requested him. I owe them all my sincere thanks.

To my family I owe a special debt. My wife Rajalakshmi Karanth contributed by listening to my loud thinking while writing the thesis, and now by going through the several drafts. My daughters, Puja and Prarthana were helpful angels. My brothers Sathyanand and Ramadas gave the moral and financial support to pursue my studies at Bangalore and Delhi. To them and to Sumitra, Suvarna, and Shrimati, I may not be able to thank adequately.

My parents have been patient with me and given me the needed understanding, concern and love through all these years. These have kept me going. I know thanking them would be belittling all that they have done. Instead, I dedicate this book to them.

February 6, 1994

G.K. KARANTH
Institute for Social and Economic Change
Bangalore 560 072

Contents

List of Tables

1

The Argument

Indian sociology has often been accused of being excessively preoccupied with institutions like caste, family, kinship and marriage, and neglecting other areas such as its economy, particularly its agrarian structure (Joshi 1975; Beteille 1974), and that they have lacked a 'development' orientation in their content (Srinivas 1979). Criticism on such a preoccupation appears to be unjust as the earlier studies were in response to the needs of the times in which they were being carried out. Indeed there is now an urgent need to reassess the continuities and discontinuities in these institutions in the context of contemporary social, economic and political processes. However, since the late 'sixties and 'seventies, both sociologists and social anthropologists have diversified their interests in studying rural societies in India[1]. Further, most of the earlier studies on Indian villages were based on field work carried out in the 'fifties and 'sixties. In the last two decades dramatic changes have occurred in rural India. These need to be captured through fresh intensive village studies, which have unfortunately gone out of academic fashion today.

This book is concerned with the nature of the emerging agrarian relations in a village in the context of some state initiated programmes such as land reforms and sericulture, and some welfare measures such as relief to the rural indebted, abolition of bonded labour system, minimum wages for agricultural labourers, etc. In examining the nature of changes occurring in agrarian relations I have used the concept of "core institutional order" as employed by Lockwood (1964), and later, by Oommen (1984). Lockwood suggests that a society may be said to have changed from one type to another only if the core institutional order in it has changed. The assumption underlying such a view is that changes in the core institutional order results in cumulative changes in the entire system. Oommen, in turn, suggests that this criterion may further be fruitfully applied to understand the changes in a particular sub-system of a society. Thus, if a village is viewed as having its own social structure, its religion, political institutions, family and kinship organisation, caste institution governing the relations between different persons, agrarian structure etc. may be identified as its sub-structures. Following Oommen (1984:16) one may identify the core elements in any one of these sub-structures.

Daniel Thorner (1976:8) uses the term 'agrarian structure' to refer to the network of relations among the various groups of persons who draw their livelihood from the soil. It is the "sum total of ways in which each operates in relation to other groups". For Joshi (1975) the study of "agrarian social structure" is primarily the study of groups connected with land. His definition is primarily in terms of relationship existing between the owners of land and the actual producers. However, these two definitions by Thorner and Joshi refer mainly to one part of agrarian social structure: agrarian relations— the relations among different groups of people or classes in the process of cultivation of land. Andre Beteille (1974:25) holds a wider view: "the study of agrarian system will centre around the problem of land and its utilisation for productive purposes". He recognises the interrelatedness of the structural elements in agriculture such as technology, work cycle, organisation of production and agrarian hierarchy. Following Beteille, it may be stated that the agrarian social structure comprises land, cultivation of which requires people, a given technology and interrelations of these dimensions. In addition, the state and the physical environment also determine the character of an agrarian social structure.

The definitions by Joshi, Beteille, and Thorner, suggest the "core" elements of agrarian social structure, although Beteille's conception is a much wider one, and hence leading to certain difficulties in selection. Joshi recognises this difficulty:

> In approaching the facts of the situation one is confronted with such a bewildering mass of details that there is a danger of getting lost without a general frame of reference for discriminating between the essential and the non-essential facts (1975:15)

Such a confusion arises because of the interrelationship between agrarian structure and wider social structure (involving the 'core' elements of other sub-structures) in shaping the particular pattern of agrarian relations. Note, for instance, Singh's (1964:34) definition which reduces agrarian relations to mean "the rights and statuses of cultivators *vis-a-vis* the state — which has become the rent collector and regulates the rights and status through legislation". While recognising the importance of the state, it is also important to take into account the nature of technology which to a large extent determines man's relations with one another, given the differentiation in ownership, control and use of land. For the purpose of the present study we may identify land, technology, and a network of relationships in the agrarian hierarchy as the *core* elements of agrarian social structure.

Transformation of agrarian social structure in India has been, by and large, a state induced process rather than being spontaneous: Thorner had observed that

> the agrarian history of India is replete with instances of efforts to change these relationships (i.e., the ways in which one operates with other groups) by law, efforts which have always fallen short of their goals or had effects other than those intended. (1976:8; parentheses added)

It is the contention of this study that measures aimed at changing the agrarian social structure or elements therein have, by and large, met with limited success in bringing about the desired change. As a result, changes in agrarian relations have been either superficial or accompanied by the persistence of the traditional elements in it, if not resulting in unintended consequences. The "core" features of agrarian social structure have persisted. I shall first demonstrate this with reference to the pre/and post-Independence periods as obtained in the available literature.

Agrarian Social Structure in India before 1947

Out of the chaos of diverse forms of land tenure, the British administration evolved three basic types of land tenure which were modelled along the lines of English manorial system and peasant type of proprietorship (Frykenberg 1977a: 8-10).

The first was *zamindari* system of land tenure, under which the zamindar held more than one village under his control. His right and control over land extended not only over people connected with its cultivation, but also a social and political authority over the people living on such land. He may have held some land for personal cultivation (*khud khast*), which he cultivated with the help of hired or attached labourers while the remaining was leased out to tenants and sub-tenants. Towards the end of British rule, in spite of the legal measures to protect the rights of tenants and to provide security to them from eviction, there existed a social situation characterised by tenants-at-will who were vulnerable to the whims and fancies of the zamindars. The second, *gramwari* land tenure, pertained mainly to the rent or revenue village as a corporate whole. Such "village community" was not the village population, rather was that segment of the elders, leaders and notables of high castes who governed the affairs of a village and faced the revenue demands made by the superior officer backed by higher political authority. The third, *ryotwari* tenure recognized the ryot as one of the leading villager and was intended to deal with him directly.

It has been argued that the introduction of both the *zamindari* and *ryotwari* settlements resulted in widespread misery for the small cultivators. The rent to be paid to the state in *ryotwari* areas and to the landlord in the *zamindari* areas were to be in cash instead of the customary rent in kind as before. As a result farmers were said to be increasingly indebted to the money-lenders. In the *ryotwari* area the climax of this situation was that the land began to be alienated from the small cultivators to the moneylenders and merchant class living in towns and cities.

The changes introduced in the then prevailing land tenure were made by the British administration with a view to further their objective of revenue maximisation and to gain support from the landed classes. New *zamindaris* were created and auctioned to the highest bidders. Revenue rates were enhanced and such *zamindars* who could not pay the enhanced demands were replaced by the new allottees. The actual cultivators were deprived of their traditional rights including the security of tenures. This process continued till 1793 when, on the basis of a permanent settlement, most of the old *zamindars* were replaced by new ones drawn from urban traders and moneylenders, who acquired wealth and influence doing business with the Company. Frykenberg points out elsewhere that there were attempts by the British to secure allegiances and loyalties of the many local elites in South India through *inam* settlements (1977b:37).

In addition to the intention of gaining legitimacy from the

landed class, the British were also guided by a notion of improvement of agriculture. It is argued that the British intention of creating a commercial agriculture in India in which the landowners would make use of the new knowledge and technology, did not result as intended and implied in the land policy (e.g., see Neale 1979:11; Mukherjee 1933:2). Neale argues that, while the British administration anticipated improvement in agriculture as a result of changes in the land policy, the *zamindar's* notion of improvement of land management was quite different. In the first place, information on improvement was not available to him. The Indian landowner was more pre-occupied with the problem of security than with improvement. His concern was primarily to gain or retain his power position for which he needed land, for, with the control over land he could control people. Elsewhere, Neale comments on the consequences of these differences between British notion of commercial agriculture and that of the Indian cultivators:

> It was long time before the commercial sensitivity of the landowners approximated the English idea of what it normally ought to be. But the restraints of obligation and of self interest never became effective... (for the Indian cultivator). The increasing commercialisation of Indian agriculture led not to a landlord-tenant alliance for progress, but to an increasing divergence of attitude and interest between landlord and tenant, and between the tenant and labourer....British policy had not led to economizing agriculture but rather to a system of monetary exploitation by the landed classes of the tenants, labourers, craftsmen, and servants of the village (Neale 1962:285; parentheses mine).

Many other scholars have pointed out the difference between expectation and actual results of the British land policy in India. Some of them have held British agrarian policy to be responsible for the deteriorating conditions in the pre-Independent Indian agrarian social structure, characterised by hardship of small holders and maldistribution of landownership (e.g., see Mukherjee 1933), and increase in the size of landless labourers and tenants-at-will (see Patel, cited in Kumar 1961:238).

There is a growing body of literature which disclaims the role of British administration being solely responsible for the state of affairs in agrarian structure. Dharma Kumar (1961 and 1962) rejects the widely held notion regarding the devastating impact of the British land policy and administration in creating an agricultural proletariat. By examining the data on caste and agricultural labour she disproves the assumption that agricultural labour was a quantitatively insignificant group when British rule began in India. Institutions like slavery and serfdom had become so complex that they appeared to have been institutions of long standing, and not a creation British policies. Likewise, Ray argues that

> the changes brought about by circulation of titles did not bring about a redistribution of land, for land was held much lower in the hierarchy of land rights and it was not until one reached the village, of mauza at the bottom of the hierarchy that people actually concerned with land were to be encountered. (1979:252)

She also rejects the popularly held notion that the lands were transferred from the old aristocracy to the hands of *banias* or other merchant classes in Bengal as a result of the *zamindari* system. She asserts that

> what actually changed hands were titles to revenue collection, not lands. By itself the change of proprietorship in *zamindaris* and taluks, the substance of which lay out in ownership of land but in dominion over villages, could not bring about a massive redistribution of agricultural lands (1979:252).

With all the efforts of British administration in restructuring the agrarian social structure in India, prior to Independence it was characterised by a high degree of concentration of landownership at the top and a large class of tenants who held no proprietary rights over the land or very little rights as tenant cultivators. Extensive tenancy and sharecropping was a dominant mode of production in which the tenants gained very little by improvement or in cultivation. The recommendations of the Commissions of Enquiry set up to examine the problems of tenants and agricultural labourers were found to be biased in favour of the landowners and *inamdars*. The legal protection given to the tenants remained only on paper (see Rao 1929; and the Report of the National Commission on Agriculture 1976, henceforth, NCA Report). There was a sizable class of landless labourers who worked on the land for wages, who were generally indebted and existed as attached labourers. The technology in agriculture remained, by and large, traditional with very little facility for irrigation. During 1949-50, just after Independence, only about 15 per cent of the total land was irrigated (Byres 1974). Use of modern agricultural inputs like fertilizers, better seeds, and pesticides was low. Under the British rule, the agrarian social structure "tended to perpetuate a backward and *medieval* type of agriculture which kept Indian economy in a state of stagnation for decades" (NCA Report 1976: 146).

Post-Independent Measures: Land Reforms

Since Independence several measures have been taken to alter the traditional agrarian structure and ensure socio-economic justice in agrarian relations. Land reforms programme may be mentioned as one such attempt which gained both political and economic importance in the Independent India for a variety of factors. During their struggle for freedom, the leaders of National Movement had mobilized peasants with the promise of land to the tiller and an egalitarian agrarian social order. The new government led by the Congress Party had to therefore fulfil their promise to the peasantry in order to gain legitimacy and support from such sections for their power (see Myrdal 1977: 222). The economic impetus for land reforms came from the view that the prevailing agrarian social structure was not suitable for agricultural growth, as it had reached a stage of stagnation. Therefore, land reforms became an essential component of the wider socio-economic development programmes which would deliver the agrarian structure from the clutches of semi-feudal production relations and thereby strengthen the essential food and raw material base for the economy (NCA Report 1976: 48-9; Joshi 1975). Land reforms gained importance, politically and socially, also because of the growing rural discontentment and frequent violent outbursts which posed a threat to the country's political and social stability.

As a first step in fulfilling the promise made to the peasants, the Congress appointed the Agrarian Reforms Committee in 1949. As it has been generally understood, the programme of land reforms could have been implemented with the aim of replacing the existing pattern of landownership either by (a) transforming the actual tillers into owner-cultivators by a large scale and radical redistribution of land among the small peasants and labourers, or (b) by inducing the landlords themselves to take up actual cultivation by hired labourers (Joshi 1975: 168). Instead the policy makers in India adopted a mid-way approach, at least in the initial stages, by abolishing the intermediaries. In the process, tenants were evicted from land much before the policies became effective.

What were the achievements of abolition of the intermediaries? It is true that it simplified the land-man-state relationship considerably. In Uttar Pradesh state alone, for instance, from about forty types of land tenure three types of uniform tenure emerged as a result of the new legislation. Official sources claim that as many as twenty million cultivators were brought into direct contact with the state as a result of abolition of intermediaries. But it also involved major difficulties particularly in identifying the real tenants, which to a large extent diluted the effectiveness and the spirit of legislation. For example, following the Bihar Abolition of Zamindars Act, 1949, there were prolonged legal battles between landowners and the state and between tenants and landowners (Sinha 1978). The *zamindars* who had controlled as much as 90 per cent of the agricultural land in the state had all the land records in their possession. The state had little knowledge as to who were the tenants or the extent of land owned by the *zamindars*. In Uttar Pradesh, where the *zamindari* system of land tenure was extensive, the loopholes in the legislations were made use of by the landowners to escape disintegration of their holdings and to evict tenants. It has been argued that however radical the abolition of *zamindari* system may appear to be, the net result was not a fundamental change in the status of cultivating farmer. Following the abolition of intermediaries, the cultivator now paid his rent to the state instead of the *zamindar* (Ladejinsky 1977: 190-1). It has also been argued that the abolition of the *zamindari* system was more beneficial to the *zamindars* than to the poor tenants as the former were able to retain an enviable and sizable portion of land for personal cultivation while also getting a handsome compensation for the rest of land. Whitcombe brings out this fact lucidly: "The impact of reform was subtle. *Zamindari*, as legal institution, was gone but its abolition had produced no miraculous transformation of the agrarian scene two decades after the passing of the Act" (1980:157). She also found that—

> Under the land reforms, all but this *sir* was acquired by government and dispersed amongst the ex-tenants, — amongst

> perhaps three times as many claimants as there were plots, who were left to their ingenuity and prospects offered by the consolidation act to make these fragments viable. The ex-zamindar, on the other hand, emerged with exclusive rights to his *sir*, and his mango-grove, and compensation for the lost *zamindari* with which to embellish them (1980:159).

It has been claimed that the path of land reforms adopted in India is one of peaceful transformation with a radical ideology. The effect, however, was not in keeping with the radical ideology of "land to the tiller". According to the literal meaning of this slogan, real beneficiaries of land reforms ought to have been the tenants and labourers who actually cultivated the land. But, agricultural labourers, by and large, have not been covered by land reforms, particularly with the half-hearted implementation of ceiling laws in different states.

While analyzing the success of land reforms a distinction has to be made between the number of cases of tenancy reported to the official agencies for settlement, and the magnitude of tenancy that existed at the time of legislation. Such a distinction is necessary because, as has been reported, of the tendency for evicting the tenants from the land in the wake of land reforms as well as settlement outside the administrative machinery. Henchette (1972) observed that in erstwhile Mysore the land economy had changed in anticipation of the land reform legislation, thus evicting the tenants from land. Sonachalam (1970) had, likewise, noted that in a district of Tamil Nadu, there had been 24,555 cases registered for settlement which was only 3.3 per cent of the estimated tenant households. A great majority of them had been settled outside the official machinery. It is anybody's guess as to who were in an advantageous position in such informal settlements. Mencher (1978) suggests that the tenants had given up their tenancies because they were unable to face the power of bigger landowners in the villages. However, it is not possible to realistically estimate the extent of tenancy, for the available land records were in such a disarray (see Manor 1980; Hill 1982; and Sinha 1978 for an account of the weakness in the official records in Karnataka, and Bihar).

A further contradiction comes from the social character of the ruling elite who were responsible for formulation and execution of the land reform programmes. The legislators and to a large extent the administrators, who were entrusted with this task came from the landed classes. As a result, not only the policies were biased in favour of the landowners through various loopholes in the legislations, but also gave enough room and time for the landowners to manipulate their interests even before the legislation became effective (see Planning Commission 1973; Mencher 1970). Large scale evictions, tampering with land records, and "redistribution of land on paper" among relatives took place in the wake of reforms. Moreover, land reforms were in the air even before the enactment which, combined with the delay in passing the laws, made it possible for the landowners to reduce the number of claimants for rights, by evicting the tenants and reducing them to the status of agricultural labourers (Ladejinsky 1977: 378; Henchette 1972; Manjappa 1984).

The bias in the early reform measures in favour of the landowners could be seen from the provision for them to reclaim land from the tenants and the grounds for evicting tenants. The National Commission on Agriculture (1976) listed as many as twenty-four different grounds on which a landowner could evict tenants under the provision of the law itself. In fact, the protection given to the tenants unleashed a contradictory social process in motion by which the tenants were subjected to mass eviction through various legal and extra-legal devices (NCA Report, 1976 : 76). Some scholars found that, not withstanding the land reforms, there remained a sizable class of land holders who were rentiers, and that during the 'sixties there had been a rise in their size.

Thus although there had been a major change by way of passing the lands to those who took active interest in cultivation (thereby reducing the size of absentee landowners), there seemed to have been little *redistribution of land from the rich to the poor* as conceived by land reforms. Tenancy abolition, for instance, in Kerala had relatively better impact in favour of the rich peasants and capitalist farmers (Herring 1980). In an evaluation of changes in agrarian structure since Independence, Joshi (1975) hypothesis that the agrarian policy of post-Independence India has succeeded in accelerating the process of disintegration of traditional agrarian system, while it is yet to substitute anything stable and viable in its place. In sum, while the programme of land reforms may have brought a change in the nature of interest in land, its lopsided implementation gave sufficient room for distortion of the stated goals. As a consequence, the class content of agrarian social structure continues to be the same as in the past. A vast majority of the tenants were evicted from land and/or reduced to landless labourers.

Impact of New Technology and Welfare Measures

What has been the impact of the introduction of new technology, and of some measures of development and welfare in changing the pattern of agrarian relations? In order to assess the impact of new technology I shall examine the case of green revolution; and with respect to the latter an examination of a set of welfare measures like the relief to the rural indebted, extension of cooperative credit to different sections of society, the abolition of bonded labour system, and provision for minimum wages for labourers. In such an examination, the main concern is to note their consequences upon the 'traditional' agrarian relations.

It has been argued that the green revolution has augmented the disparities in income and living standards between the rural rich and poor (e.g., Frankel 1971; Chaudhari 1974). Some scholars have also indicated that as a result of green revolution wide ranging changes in the agrarian social structure have occurred. For example, Aggarwal (1975) identified three categories of changes in Punjab. They are transformation of subsistence oriented peasant economy into commercial and profit motivated farm economy and traditional technology centred around bullock cart into industrial technology; and breaking down of the isolation of peasant community as new links with civilisation were being forged. Some of these observations need to be examined for the country as a whole while they may be true of pockets where green revolution was concentrated. Use of bullockcart and cattle continues to be dominant in the agrarian system of the country in spite of the new 'industrial technology'. Moreover, the notion of isolation of the peasant economy seems to be a relic of literature which assumed Indian villages to be self-sufficient microcosms.[2]

Frankel (1971) finds erosion of traditional attitudes of mutual dependence in the relations between the landowners and labourers, increased tension as a result of disparities in income distribution (for similar observation, see also: Brown 1970; Singh and Singh 1975; Sen 1970; Alexander 1975; Lakshminarayana 1977). However, such unilinear explanations about the erosion of traditional relations and increasing agrarian unrest as a consequence of green revolution have been found to be inadequate to explain the rural unrest (Oommen 1971a and 1975). Joshi (1975) has argued that the change in technology has not been a sufficient cause for bringing about changes in the 'traditional' agrarian relations.

Although green revolution appears to have contributed to an increase in their wages, the increase in prices has kept the real income of labour household static (Lakshminarayana 1977). Some scholars have pointed out that even in the erstwhile I.A.D.P districts, in spite of agricultural modernisation and development, wages have fallen. Thus, at the end of the 'sixties more than seventy per cent of the rural population in India were below the barest minimum accepted level of poverty.

The frequently made reference to the disappearance or erosion of traditional features of agrarian relations by some scholars, however, fails to specify the content of the 'traditionality' (e.g., Brown, Singh and Singh, Sen, Alexander, and notably Frankel and Aggarwal, as cited above). What is even more curious is the fact that they do refer to the existence and persistence of some traditional (by implication) features of agrarian relations although their conclusions are to the effect that the 'traditional agrarian relations are being replaced by impersonal and monetised relations'.

This throws light on the approach prevailing in understanding the process of social change in rural India, which is based on the assumption that traditional and modern features of a phenomena cannot coexist. Gusfield (1967), though in a different sphere, disagrees with the assumption of traditional and modern forms being conflicting and mutually exclusive systems. Rudolf and Rudolf (1967), likewise, hold the view that radical contradiction assumed to lie in the character of modern and traditional forms stems from the misconception about the 'traditionality' of the traditional societies and 'modernity' of the modern societies, and a misapprehension of the relationship between the two.[3]

One of the sources of confusion in the debate on modernity and tradition in the analysis of social change in countries like India is a failure to distinguish the "core" institutional order in a given society as referred to by Lockwood (1964) and Oommen (1977). Oommen argues that the perspective taking into account of 'core institutional order' will help us to understand not only the contradictions involved in the discontinuities but also the reason for the co-existence of curious mixtures of modern and traditional elements of social structures and values in a given situation. (1977:90).

In order to examine the nature of transformation of 'traditional' agrarian relations into market relations of the modern type, we need to identify the 'core' elements of both traditional and modern agrarian relations which most of the studies cited earlier rarely specify. These studies do point out that the majority of labourers have been drawn from the lower castes, particularly the former untouchable ones, and that the landownership has, by and large, continued to concentrate in the hands of the dominant peasant castes, who had been traditionally the big landowners. Further, the mode of payment to the labourers becoming increasingly in cash rather than in kind is also referred to as being a sign of monetisation of agrarian relations. But they rarely refer to the details regarding the mode of wage payment in the past and for different operations, while tending to assume that it has always been in kind in the past. While demonstrating the 'breakdown of traditional interdependence' of people, they also emphasise the continued indebtedness of the rural poor, particularly the landless labourers and scheduled castes, to the rich farmers and moneylenders. Interdependence of different sections in rural areas in the past did involve monetary transactions of borrowing and lending, and working for others for wage payment.

With regard to the findings on customary services under the *jajmani* relations, it is paradoxical when one finds the studies concluding that as a result of new technology the traditional *jajmani* relations have disappeared. Aggarwal (1971), for example, notes that as a result of green revolution "the first casualty seems to be *jajmani* relations". In the same essay he also recognises the fact that there is a differential adoption of new technology among the big and small farmers. The question that comes to mind then is, why do the small farmers discontinue *jajmani* relations with artisans and other serving castes? Why do the *kameens*, in turn, refrain from serving the small farmers whose adoption to new technology has been relatively less than the bigger farmers, and who have remained, by and large, subsistence farmers? Once again, the source of such contradictory accounts of social change lies in the unilinear explanations offered for changes in agrarian relations whether in the case of green revolution or legislative measures. However, it may be recalled here that there have been a few studies reporting the continuity of the traditional *jajmani* relations despite the cash income to the farmers (e.g., Harper 1959) and adoption of new technology (Harris 1980).[4]

I shall now proceed to examine the impact of some of the welfare measures related to agrarian relations. There have been measures to maintain a minimum level of agricultural wages throughout the country. The minimum wages fixed have varied widely from state to state, and most of the states had not taken enough steps to enforce them. According to the National Committee on Labour (1960) in almost all the states the Minimum Wages Act had remained a dead letter. Six years later, the Rural Labour Enquiry reported that only two percent of agricultural labourers were aware of the Minimum Wages Act. Maharashtra, for example, claiming to have taken into account the minimum needs of the people in fixing the wages, decided to fix the rate as low as Rs. 4.00 to Rs. 5.50 per day, which however, was found to be even lower in practice (Correspondent, *E.P.W.* 1980).

Fixation of wages to be paid for labourers does not in itself ensure that it shall be followed. The mechanism of wage determination in the villages falls into account many other factors than mere supply and demand of labour, and the minimum needs of the people. Commenting on the models that are put forth to explain the dynamics of the rural labour market (e.g., Lal 1976), Jose (1978) points out that such models fall short of adequacy to present a complete picture. Often they omit such aspects as the proportion of women and child labourers in the rural labour force and fluctuation in seasonal demand for labour. To assert such relations between agricultural output and income levels of agricultural labourers, one has to go beyond the realm of figures; for, often many of these phenomenal factors which we find in rural areas such as caste, systems of bonded labour, and indebtedness, which are far more important in the process of wage determination, fail to be caught by sophisticated models designed to explain the behaviour of wages (Jose 1978:A-20). As a result of such local factors, mere legislation is not sufficient to bring about any considerable improvement in the wage level or living conditions of agricultural labourers.

With regard to credit facilities, measures had been taken to ensure equal access to all sections of the rural population through commercial banks, credit cooperatives and agencies like the then Small Farmers Development Agency, and the like. To free the rural poor from the clutches of rural moneylender relief measures were initiated. Not withstanding all these, informal credit at arbitrary rates of interest are found to be persisting. Although during the mid 70s the Reserve Bank of India had issued directives to the credit institutions to provide a minimum of 20 per cent of all their advances in the agricultural sector to the small holders, it was found that the small farmers themselves had not been availing of credit facilities. The economically vulnerable sections in the rural areas continue to find it difficult to have access to the institutional credit. Credit policies still show bias in favour of the wealthy landowners who have been able to corner a major chunk of institutional resources for themselves (see Jodha 1971). This is because of the linking of institutional credit to the 'credit-worthiness' of the farmers. The Rural Labour Enquiry (1974-75) reported that the share of the institutional agencies in the total debt of landless labour households is negligible. The explanation offered by the institutional agencies has been that they cannot lend more to the weaker sections of the rural population because their borrowings are essentially for consumption, and the recovery of such loans becomes doubtful.[5] Indeed, the report of the Enquiry finds an increase in the share of non-institutional sources in the borrowing of labour households over a decade, which rose from 30.6 per cent in 1964-65, to 48 per cent in 1974-75. On the other hand, the share of cooperative credit to labour households during this period had increased only negligibly from 5.1 percent to 5.3 per cent, while the corresponding figures for Scheduled Castes registered a decline from 4.7 per cent to 3.8 per cent (Rural Labour Enquiry, cited in Editorial, *E.P.W.* 1979). It has been pointed out that, "the prevailing system of land tenure tended to be reproduced as a differential right to institutional credit, often preserving the very distinction which was the core of the traditional agrarian system" (Herring 1977:121). Consequently, it may be seen that the traditional dependence of labourers upon landowners and moneylenders persists in rural areas, despite the changes in technology as discussed earlier.

There is a growing body of literature which is presenting a new picture about the role of moneylenders in the rural areas. Rudra (1975), for instance, calls for reconsideration of the prevailing generalisations pertaining to modes of loan transactions. In his study of Bengal villages he observed different kinds of lending, by the landowners, which are not necessarily usurious. The consumption loans to the farm servants were often interest-free, which shows the persistence of traditional interdependence of the landowners and labourers. Michie (1978) observes that although the moneylenders (*banias*) have a bad reputation, they have been recognised to be performing vital services as a source of credit for production and consumption, and that they are not "all take; they do give as well". In the absence of an effective credit facility any attempt to close the doors of local moneylenders to the borrowers particularly to the small and marginal farmers and landless labourers would be an ill advised measure when the vacuum created by them is left untouched.

Poly Hill (1982:216) questions the current opposition to the village generated credit, and points out that among other reasons, such opposition is inherited from the colonial obsession with debt. She argues that the real urban exploitation... consists in the endeavour to suppress village-generated credit, which is all that is available to poorer people, while providing no one, not even the rich, with anything like sufficient to take its place.

Some of the factors responsible for the continuity of traditional credit system in the village stem from not merely the inadequate credit facilities, but also the bureaucratic requirements of the credit for granting a loan, and the nature and composition of the officials in such institutions. Oommen (1974:44-6) points out that the cooperatives continue to be dominated by an exclusive group of leaders, and that the rules, which are laid down are seldom strictly followed. He lists non-repayment of loans, corruption by the board of directors, elaborate procedures that delay the sanction of loans etc. as factors affecting the cooperative credit societies in Kerala. The social humiliation caused by the recovery cell of the cooperatives and land banks, further discourages the farmers from approaching the formal institutions of credit while continuing to approach the local moneylenders as before. The amount of bribe that has to be paid at various levels of officers often works out to be more than the interest paid on a private loan (Sivakumar 1978). This, to a great extent, explains why there is a major share of credit raised from informal sources, despite the efforts by the state to replace them with cooperative and banking institutions.

There is enough evidence to suggest that the dependence of artisans and service castes upon the landowners has increased, if not in their occupational relations for a livelihood, at least, as agricultural labourers and for credit. The possibility of such an increase in dependence comes from the fact that they too, like agricultural labourers, are unable to get institutional credit. According to the All India Debt and Investment Survey Report for 1971-72, the share of the institutional agencies in the debts of artisan households was only 5 per cent while 70 per cent came from the landlords, moneylenders and traders.[6] This only shows the continuity of dependence of artisans and service castes on landowners for credit, if not in the network of *jajmani* relations.

Further, with regard to the coming of new technology, and credit relations, Dasgupta (1977:245) observes that most of the finances for purchase of new inputs has come from the farmers themselves while the reliance upon institutional finance is less than one-tenth of the total borrowings. He attributes this to the rigidity of bureaucratic formalities faced by the poor landless and tenants in gaining access to institutional credit. Hence, the continued dependence on the traditional sources of credit.

This continuity has significant implications for understanding the interrelations between the different sections of rural society. The bigger landowners and moneylenders, therefore, are found to be the major sources of credit for the rural poor. The chief motive for the moneylenders is to create social obligations than to gain direct profit from interest payments (Roth 1979). Such social obligations facilitates the moneylenders to establish claims or perpetuate leadership in the traditional setting.[7]

It has been often argued that in the wake of industrialisation and urbanisation, rural/traditional societies are being modernized. Going as far back as 1893, Marx had observed for India that modern industry, resulting from the railway system will dissolve the hereditary divisions of labour, upon which rest the Indian castes, those decisive impediments to Indian progress and Indian power (Mary and Engels 1976:497). There is no denying the fact that over the century since Marx wrote on India, such association between caste and occupation has been diversified. Even so, there is enough evidence to point out the persistence of inter-caste association under the *jajmani* system despite the coming of industry, adoption of a new technology, and the consequent economic development (see, e.g., Thakurata 1976; Harris 1980; Epstein 1962, 1967 and 1973).

Similarly, continuity is to be found in some aspects of the traditional landowner-labour relations also, particularly the system of bonded labour. Bonded labour system in agriculture came into limelight both politically and academically during the mid 70s. Under the 20-Point Programme, this system was sought to be abolished. Accordingly, many states enacted legislations to abolish the system, and claimed to have identified, released and rehabilitated scores of them. Even after a few years of such claims, many studies reported the persistence of the system under disguised forms, if not as before (Joshi 1975; Dalwai 1977; Shah 1978; and Kulkarni 1979). Indeed the increased demand for agricultural labour, which in turn is a consequence of intensive cultivation of land following the adoption of new technology, has been one of the causes for the persistence of a need for attached labour (Joshi 1975).

Change and Continuity: The Problem for Study

Two broad types of conclusions emerge from the foregoing analysis of literature on agrarian reforms and rural social change. The first of these relate to "change" from the traditional, customary and interdependent relationships between different agrarian categories. According to this perspective, there has been change in the social and economic status of labourers; a breakdown of interdependent relationship including *jajmani* relations; and in the place of such traditional relations there has emerged a monetised, impersonal, bargaining relationship. In other words, agrarian relations are found to be changing from the "traditional" to commercial and impersonal relations.

The second refers to distortions in the implementation of many agrarian reform measures like land reforms, introduction of new technology, credit facilities, and measures to modernize agrarian relations by abolishing bonded labour system and rural usury. Consequently, the traditional landowning classes have retained their social and economic dominance, and that there is a persistence of traditional forms of dependence of the weaker sections upon the landowners, attachment of labour, and relatively minimal changes in the wage structure.

It may be pointed out that, following Lockwood (1964) and Oommen (1967 and 1977), one of the sources of confusion in analyzing rural social change in India stems from the failure to distinguish the 'core' institutional order from the total structure. Furthermore, following Gusfield (1967) and Rudolf and Rudolf (1967), in analyzing social change it is necessary to view tradition and modernity as coexisting possibilities and not poles apart. The present work is an attempt to demonstrate how 'changes' in the agrarian relations have been accompanied by the 'continuity' of the past or with persistence of tradition. Such a dual process needs to be recognised as distinct from either "changing" or "change-resistant" process, for measures of agrarian reforms in India have not addressed themselves to a total or substantial change. The present work also attempts to point out that the observed "continuity" is due to the gaps created by the changed situations. Therefore, until the vacuum is filled up by other measures such 'traditional' or pre-change features will continue.

It may be added here that the above perspective does not deny changes in agrarian relations, nor deny the potential of change producing ability for measures like land reforms, new technology, or welfare measures to better the life and working conditions of the labourers. It is relevant to note that legislation is not a sufficient, though, it may be necessary, mechanism to institutionalize social change — particularly when change is sought to be brought about in favour of the poor, in a society characterised by extreme concentration of wealth, power, privileges...(Oommen 1975:1570)

It is intended in this work to show how traditional features of agrarian relations continue to adapt themselves to new conditions in rural settings. It is said that Indian social structure is capable of accommodating or adapting itself to new values and patterns of behaviour (Eisenstadt 1970). It is this structural adaptability which explains coexistence of traditional forms of relations in the new and changed situations (see also, Mandelbaum 1972). Moreover, the traditional forms of relations have some functions to perform, by which it reproduces itself.

As land constitutes one of the core elements of the agrarian social structure, an attempt has been made to examine the nature of changes, if any, in ownership, control and use of land as a result of land reforms. Secondly, the study aims at examining the impact of the introduction of a commercial crop, namely, mulberry and the associated rearing of silkworms (i.e., sericulture), upon the existing agrarian relations. In this context, it addresses itself to examine changes in the nature of inter-caste dependence of different sections of agrarian population. Thirdly, an attempt has been made to examine the nature of agricultural labour arrangements in the light of several measures such as debt relief, abolition of bonded labour and provision of minimum wages for agricultural labourers.

Methodology:

The study was carried out in the village Rajapura[8], Magadi taluk of Bangalore District, Karnataka State in South India. The choice of the village was on the basis of manageability of its size for a single researcher for an intensive study. At the time of the commencement of my study, a team of sociologists and social anthro-pologists were also engaged in studying the same taluk[9] with a micro-macro approach, and it was suggested that more intensive studies in different villages with different problems for research would enrich the sociological

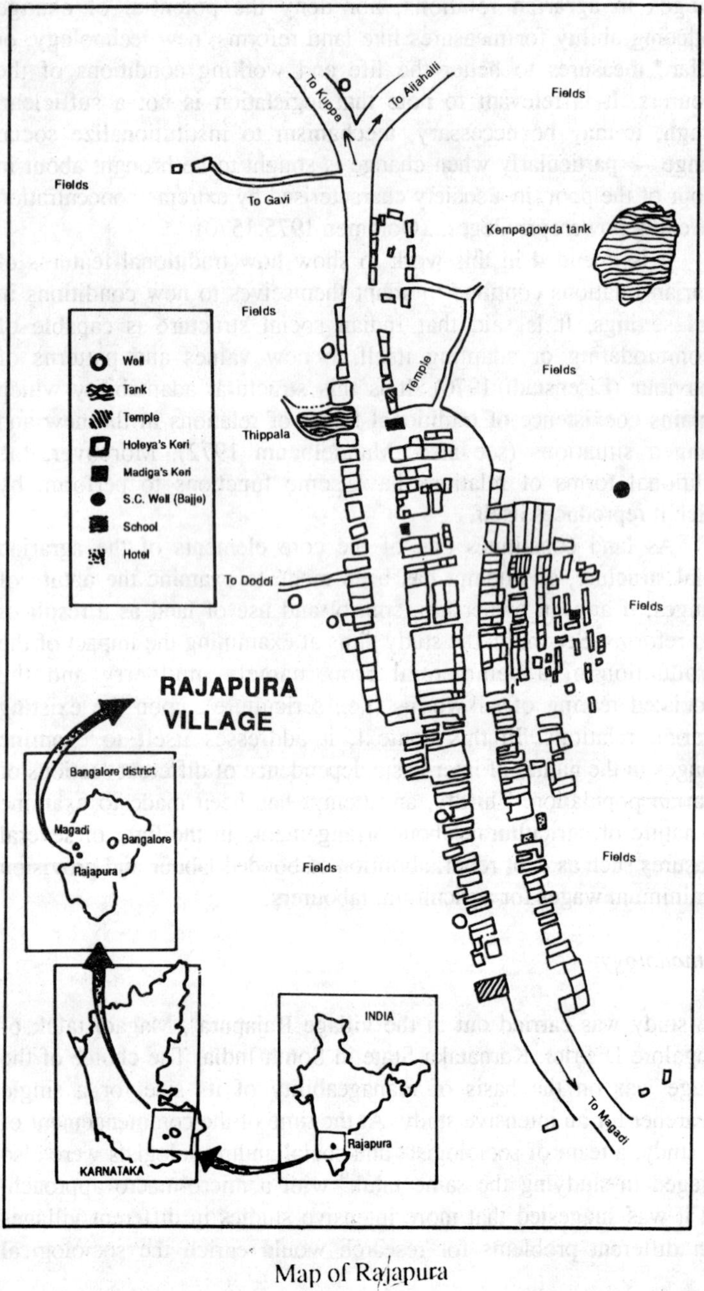

Map of Rajapura

insight into the understanding of local and supra-local factors of change and planned development.

Initially, a list of villages in Magadi taluk with 650 to 800 population, and households with 130 to 150 was prepared. I visited these villages for a preliminary examination, and Rajapura was chosen for the study since it was relatively interior with no vehicular transportation available at that time, and which had widespread sericulture and dry farming. I lived in the village from September 1978 till May 1979, again from June 1979 to August 1979 and December 1979 to April 1980. Participant observation was supplemented by data gathered through a census study of all households and informal interviews. Village land records and files in the department of sericulture were made available to me at the village, taluk and district levels. Officials at these levels were also interviewed, and with special permission, I was permitted to observe the proceedings of the Land Tribunals in Magadi Revenue Office, during the course of which I could attend several "spot inspections" by the officials and members of the tribunal in disputed cases. Several prominent persons from the neighbouring villages were also interviewed in order to ascertain the validity of some information gathered about the village (See Map of Rajapura).

The Field Setting :

The study was carried out in village Rajapura, an interior village in Magadi taluk of Bangalore Rural District, Karnataka. It is situated to the west of Bangalore city, the district headquarters, at a distance of about 60 kilometres. Magadi town, the taluk headquarters, is to its north-east at a distance of about 9 kilometres. Its neighbouring villages are Shambu Halli to its west, Palya to its north, Kuppe and Doddi to its east and south-east respectively. Adjacent to it on the south-east, between Kuppe and Rajapura, is an uninhabited village Ramapalya, a former *inam* village. Nearly four-fifths of 165 acres of this village is now under the control and ownership of farmers in Rajapura.[10] Bordering on Rajapura's south and south-west is the state forest of Sidda Betta. All these villages are connected by foot-paths, with the exception of Kuppe and Palya. A *kuchcha* road passes through Rajapura, Kuppe and Palya, which joins the main road between Ramanagaram and Magadi at a distance of about 6 kilometers from Rajapura. Doddi, which is a hamlet of Kuppe, houses the office of the

TABLE 1.1 : Households, Population and Land Ownership of Different Castes in 1979

Sl. No.	Castes	Population Male	Population Female	Population Total	House-holds	Aver-age size of House-hold	Total Land owned (acres)	Average Land owned (per HH)	Average land per capita	Land-less House-holds	% Land-less HH	Male work-ers	Female work-ers	Total work-ers	Percent-age workers in the population
1.	Lingayat	10	4	14	3	4.7	16.00	5.33	1.143	2	66.7	7	3	10	71.4
2.	Okkaliga	336	339	675	108	6.3	659.02	6.10	0.976	14	13.0	198	217	415	61.5
3.	Kumbara	4	5	9	2	4.5	2.00	1.00	0.22	—	—	2	3	5	55.6
4.	Madivala	5	5	10	3	3.3	—	—	—	3	100.0	3	5	8	80.0
5.	Bestha	1	1	2	1	2.0	—	—	—	1	100.0	1	1	2	100.0
6.	Achari	13	6	19	3	6.3	10.10	3.37	0.53	2	66.7	3	4	7	36.8
7.	Maratha	9	10	19	3	6.3	18.38	6.12	0.967	—	—	6	8	14	71.7
8.	Lambani	21	21	42	14	3.0	24.28	1.73	0.578	4	28.6	14	17	31	73.9
9.	Korama	9	9	18	3	6	8.25	2.75	0.46	—	—	5	6	11	61.1
10.	Oddas	17	13	30	7	4.3	10.00	1.43	0.33	2	28.6	11	8	19	63.3
11.	Holeya	28	37	65	14	4.6	49.32	3.52	0.759	2	14.3	22	26	48	71.8
12.	Madiga	51	41	92	22	4.2	29.75	1.35	0.323	13	59.1	37	30	67	72.8
	Total	504	491	995	183	5.4	827.10	4.52	0.831	43	23.5	309	328	637	64.0

Mandal panchayat (prior to 1986, the village panchayat) of which Rajapura is one constituency. People of Rajapura commute quite often to Doddi, for, it also houses a higher secondary school, post office, health centre, veterinary clinic, and more importantly, the marketing centre and range office of the department of sericulture. Prior to 1984 when there was no regular bus service for people of Rajapura, they had to walk up to Doddi to take a bus or use the *kuchcha* road to Magadi. During my study one bus used to ply between Ramanagaram and Magadi via Rajapura and several other interior villages on this *kuchcha* road, but this was regular only during the season of *avarekai* (*dolichos lab ab*) when many farmers travelled to Magadi, Ramanagaram or to Bangalore city to sell their daily harvest of this and other vegetables.

During 1978-79, when this study was carried out, Rajapura had a population of 995 persons living in 183 households. The population was distributed among 12 different castes (see Table 1.1). Okkaligas being to the numerically, economically and politically dominant caste, which is the case in the region also. Members of all castes in the village engaged in agriculture, although some of them had occupations traditionally associated with their caste. The village has an area of 964 acres (1971 census) of which officially only 521 acres of land is cultivable.

Land is mainly dry. There is hardly a stretch that is free of rocks and boulders thus rendering uncultivable much of the land otherwise classified as 'cultivable'. However, such plots serve as grazing ground, or for putting up a haystack, manure pit, etc. The lone tank with an extent of about seven acres hardly irrigates three acres of land. Much of the tank's land not under water has been encroached by the farmers with holding around it, for the tank is never full. The village faces shortage of water for agriculture and drinking.[11] A tract of land to the west of the village, stretching from north to south is comparatively better endowed with underground water resources. Consequently, land in this tract has been fragmented much more than land elsewhere. Farmers have sunk wells on their small holdings in this tract, most of which are crude and go dry during summer months. They are fitted with manually operated lifts (*yeta*). There were about 20 electric and three diesel run irrigation pump sets (IP sets). Usually a plot of less than an acre to one acre is converted into garden land[12] in which a well is sunk. Here mulberry, coconut, irrigated paddy and ragi, and some vegetables are grown. If water is abundantly available sugarcane is also grown.

The main crops grown are ragi, mulberry, pulses, groundnut, and paddy. Ragi forms the staple diet of the people, occasionally supplemented with rice if they can afford it. Dry cultivation of ragi provides for mixed cropping such as mustard, black-til, castor seeds, field-beans (*avarekai*), *jowar*, and horse-gram. Only rarely *averekai, til*, mustard, etc., are grown as a separate crop. With the exception of some, all these are grown for domestic consumption. Jowar is grown mainly to feed the draught animals. Farmers sell small quantities of foodgrains, but the bulk of their cash income is through sericulture[13] and sheep rearing. Sericulture is a major occupation of nearly all the people and draws all sections of the population in its various activities. Silkworms are fed with mulberry leaves, and the cocoons are sold for 'seed' purposes in Doddi. A 'crop' of silk cocoons takes about 45 to 60 days (depending upon the season and care given). A farmer can raise six to seven crops a year, whereas a single crop of ragi requires nine to ten months a year.

Total land area under the control and ownership of people in the village is 827.03 acres, which includes holdings in Ramapalya. Eight households owned land outside Rajapura and Ramapalya, but only one of them personally cultivated such land. Okkaligas as a whole owned as much as four-fifths of the land (see Table 1.1). Average land-holding per *land owning* household is 5.35 acres. There are 43 landless households, a majority among them being non-Okkaligas. Quantitatively and qualitatively landowning pattern show a positive relation with caste status. Okkaligas as a whole own all the land under tank irrigation with the exception of a tiny plot of 0.20 acres.[14] Further, one finds that the 'upper' castes have better situated land than the other castes. The upper castes (particularly Lingayat, Okkaliga, and Maratha) have their lands situated around the village settlement. Such lands, referred and valued as *ooru pakkada jameenu*[15] has several advantages, such as easy accessibility, watch and ward, willingness of the labourers to work, etc. Even though several farmers owning land of this type are small and marginal farmers, they are referred to as *kula*[16], by virtue of the value attached to such lands. Okkaligas also own almost all the land lying next to the tank and those lying next to a small stream flowing on the southern border of the village. On the other hand, most of the landowning Kumbara, Korama, Lambani, Odda, Holeya and Madiga households have come to own and through periodic land grants by the government. These lands are not only situated far from the village site, but also are of low quality. In fact, these lands were those which were not claimed by people owing to the rocks and boulders in them. Thus, many landowning lower castes are "landowners" by definition while the lands they own are generally undulated and are difficult to convert them into good and fertile holdings. As a result, many such landowners have to seek patronage in terms of work, loans, and agricultural implements from the Okkaliga landowners, which, to a great extent, affect the former's social status. Very often such landowners work for other landowners either for a wage or without wages in return for the help they may secure from the latter.

Over the past 40 years, land transfers in the village have been progressively in favour of Okkaligas. A landless Okkaliga has a much better opportunity to own land by making use of the power, and economic status of many of his caste and kinsmen. Okkaligas are also more successful in getting land on tenancy and sharecropping, or on mortgage of land belonging to others.

The distribution of land shows a predominance of marginal farmers (less than 2.5 acres) and small farmers (between 2.5 and 5.00 acres), who constitute 59.0 per cent of all landholding households. Thus, 26.6 per cent of land is owned by 45.4 per cent of all households (See Table 1.2). The medium size farmers (i.e., with 5 to 10 acres) form 19.7 per cent of all households, and control 30.4 per cent of land. The big and very big farmers constitute only 11.5 per cent and control as much as 30.13 per cent of land. The landholding pattern in terms of

TABLE 1.2: Land Distribution in Rajapura

Sl. No.	Size Class	No. House holds	Per cent	Area* (in acres)	Per cent
1.	Landless	43	23.5	—	—
2.	Marginal farmers (0.01 to 2.50)	41	22.4	65.29	7.89
3.	Small Farmers (2.51 to 5.00)	42	23.0	154.86	18.72
4.	Medium Farmers (5.01 to 10.00)	36	19.7	251.05	30.35
5.	Big Farmers (10.01 to 20.00)	16	8.2	226.37	27.37
6.	Very big farmers (20.01 and more)	5	2.7	129.53	15.67
	Total	183	100.0	827.10	100.00

caste (Table 1.3) presents an interesting feature. Out of the 43 landless households, Okkaligas are only 13.0 per cent. All the Washerman and fisherman castes are landless while among Lingayats and Madigas more than 50 per cent are landless. The bigger size groups may be seen to be dominated by the three ritually superior castes: Okkaliga, Maratha and Lingayat. All the very big farmers are Okkaligas, constituting more than 80 per cent in the class. Even the lone Lingayat who owned 16 acres, which was the biggest landholding during 1978-79, recently sold his lands to two Okkaliga farmers.

Sharecropping and Tenancy :

The practice of sharecropping and tenancy cultivation continue to exist, which operates in an informal and clandestine manner to circumvent the law.[17] One of the major incentives for the present day arrangements of sharecropping and tenancy for the landowners is to ensure a *labour bank* throughout the year. Sharecropping is to be found mainly for cultivation of mulberry (and rearing of silk worms), vegetables, maize, sugarcane, while tenancy is more in common for food crops like ragi, horse-gram, groundnuts etc., which are grown as dry crops. The absentee landowners, who have now migrated to other places on employment, generally lease out their land on sharecropping or tenancy basis. Brothers residing in the village also may cultivate such lands on crop-sharing basis, although it is a rare event for brothers to be on cordial terms even after the division of the ancestral property. In most cases of tenancy and sharecropping, the terms and conditions are not always specified and is not given importance. Such flexibility occurs when the two farmers are kinsmen or have a relation of exchange labour. In some cases, the bonded labourers are also given a plot of land on crop sharing basis, although the landowner may not always insist on his receiving his share. This is more so if the labourer agrees to continue to be in service the following year also.

During 1978-89 there were 77 cases of tenancy and/or sharecropping. A small section of tenants and sharecroppers were landless among whom a higher proportion were for both sericulture and food crops (53.9 per cent). There were nine cases of bonded labourers getting a plot of land to cultivate on crop-sharing basis from their masters. Apart from the landless, the small and marginal farmers as well as the medium-size farmers also have sizable proportion of tenants and sharecroppers among them. It is important to note here

TABLE 1.3. Caste and Land Distribution

Caste	Landless	Per cent of Landless (of each caste's total)	Marginal Farmers (0.01 to 2.50)		Small Farmers (2.51 to 5.00)		Medium Farmers (5.01 to 10.00)		Big Farmers (10.01 to 20.00)		Very big Farmers (20.01 & above)		Total	
			No.	Area	No.	Area	No.	Area	No.	Area	No.	Area	No.	Area
Lingayat	2	(66.7)	—	—	—	—	—	—	1	16.00	—	—	3	16.00
Okkaliga	14	(13.0)	20	32.81	27	100.83	28	195.58	14	200.27	5	129.53	108	659.02
Achari	—	2	(66.7)	—	—	—	—	—	1	10.10	—	—	3	10.10
Maratha	—	—	—	—	2	8.38	1	10.10	—	—	—	—	3	18.38
Lambani	4	(28.6)	7	11.38	2	6.90	1	6.00	—	—	—	—	14	24.28
Oddai	2	(28.6)	4	4.50	—	—	1	5.50	—	—	—	—	7	10.00
Holeya	2	(14.3)	3	4.35	5	17.50	4	27.47	—	—	—	—	14	49.32
Madiga	13	(59.1)	3	6.00	5	17.25	1	6.50	—	—	—	—	22	29.75
Others*	4	(44.4)	4	6.25	1	4.00	—	—	—	—	—	—	9	10.25
Total	43	(23.3)	41	65.29	42	154.86	36	251.05	16	226.37	5	129.53	183	827.10

*'Others' include 2 Kumbara; 3 Madivala; one Bestha; and 3 Korama caste households.

that more than half the tenants and sharecroppers engage in cultivating other's land for food crops only.

Agricultural Labour :

Almost every landowner in Rajapura complains of shortage of labourers, particularly during the peak seasons. Shortage of labour occurs because of at least two reasons. First, the forest in the neighbourhood provides an opportunity for many landless and landowning households to gather firewood for both domestic use and for sale in the town. This practice is widespread among the Lambanis, Holeyas, and Madigas, who constituted the bulk of agricultural labourers in the past. For at least three days in a week now men, women and children of these castes engage in forage and travelling to Magadi with head-loads of firewood. Secondly, many of the developmental activities undertaken by the government withdraws labourers from labour force, either temporarily or permanently. This has an impact in different ways. In the first place, many marginal farmers and small farmers have obtained medium and short term loans from the credit cooperative society at Doddi to purchase sheep, cows, 23 etc. As a result, many of them are engaged in grazing the animals instead of seeking employment for wages. Secondly, work undertaken under the rural employment assurance scheme also took away labourers from agriculture. Further, the scheme 'old age pension' paid to persons above 60 years of age without property and supporting family members has also contributed for the withdrawal of many persons from the labour force, at least partially. While noting this as a factor I must hasten to add that many recipients of the scheme are not, as per rules, eligible to receive such pensions. However, many a landless and even marginal farmers, particularly female members who receive such pensions in these households, do not work as labourers although they did so prior to receiving pension. In addition to these factors, there is another way by which shortage of labourers occur. As pointed out earlier, many landless households now engage in sharecropping and tenancy cultivation. Consequently, they keep away from working for wages throughout the year, and when they do work, it is usually for the landowner with whom they sharecrop. Indeed many landowners engage in leasing out their land or sharecrop with a clear intention of ensuring a labour bank from such tenants and sharecroppers.

The agricultural labour households as conventionally understood, that is, those who derive their livelihood predominantly from agricultural wages, are too small in Rajapura. Indeed, for a meaningful discussion of agricultural labour and landowner relations, it is useful to identify agricultural *labourers* than households. Besides there being a small number of agricultural labour households, a greater part of the labour supply is from the landowning households. A majority of such labourers come from the landowning households not keeping with their landowning status. Further, many sharecropper or tenant cultivators work as labourers for the landowners, for whom income from wages is minimal while the contribution to the village's agricultural work force from them is quite significant.

As may be expected, there is a close association between one's caste and participation in wage labour. The dominant Okkaliga caste, thus, has the least number of households having persons working as agricultural labourers, while among the former untouchable castes — Holeya and Madiga — as also among the Lambanis and Oddas (all Scheduled Castes in Karnataka) this proportion is much higher.

It is relevant here to discuss briefly the conception of working day, and further distinguish different types of labour arrangements.[18] For the landowners and labourers alike, the working day is divided into two halves: forenoon and afternoon. Usually one half is devoted for grazing the animals and the other for agricultural operations, marketing, visits to the barber, temples and so on. The sequence and purpose of this division varies from household to household, according to seasons, and the nature and manner in which the work is to be accomplished (that is, with domestic, hired, or exchange labour). Those with enough working members in the households may perform these tasks simultaneously, or interchangeably in terms of what work is done during the two halves of the day. For the labourers also, the entire day may be a working day or either the first or second half for wage labour which depends upon their age, the size of land owned by them, type of farming, their other activities like firewood gathering, grazing, sharecropping etc. However, the division of the working day into two halves is not applicable to the bonded labourers, for they work throughout the day and in all seasons of the year.

There are four major types of labour arrangements made in the cultivation of land and in other agricultural occupations. They are:

(1) Wage labour
 (a) casual labour
 (b) contract labour
(2) Seasonal unfree labour
(3) Bonded labour
(4) Exchange labour

Casual labourers are those hired on a day to day basis, paid at the end of day's work. The contract labourers are usually a group of labourers who undertake to complete a given task for a specified amount of wages (*cooli*). The seasonal unfree labourers are those who work without wages, but work to repay an interest on a loan taken by them. The bonded labourers are those who work for an annual wage received in advance, with or without interest levied. There are two forms of it, one in which a worker works off the interest while the principal is to be repaid. In the other form, the worker works off the principal borrowed, and only if he absents from work does the borrowing attracts an interest. The exchange labour system is by far the most important source of labour supply to the farmers in Rajapura. It is a mutual pooling of labour for working on each others' fields. In doing so the landowners of all size, class and caste take part. This arrangement is made for operations requiring group labour such as weeding and harvesting.

Notes

1 For a discussion of sociology in India in recent times, see Rao 1979; Oommen 1984; Oommen and Mukherjee 1987.
2. See, Srinivas and Shah 1960.
3. See also, Gould (1970) who holds an opposing view.
4. See also, Karanth 1987.
5. Whatever the political character of the *loan-melas* by the commercial banks during 1983-84 and later, the objection for such mass loan-melas from the banking industry has not been any significantly different from this. See, e.g., Murthy *et.al.* (1984).
6. Cited in Editorial, *Economic and Political Weekly* (1978, pp.1007-8)
7. For example, see Beals (1970) who notes of how the income from industrial employment was being used for moneylending in the village so as to gain status in the traditional leadership settings.
8. The name Rajapura is a pseudonym used to conceal the identity of persons who may figure in this work.
9. see, Epstein *et.al* 1983.
10. Land in Ramapalaya is considered as part of Rajapura's land, since most of it is under the control of the farmers of Rajapura.

11. Few bore-wells have now been sunk easing the drinking water situation. Since 1989, a mini-drinking water scheme has been implemented though the scheme has failed to meet the needs of the people. During January 1994 there was one borewell which was catering to the needs of the people.
12. Such conversion of dry into garden land is not usually found to be notified to the revenue authorities, for it attracts a higher rate of land revenue.
13 As an agro-based industry, sericulture involves three major activities: growing mulberry, rearing of worms, and reeling. For a very useful social anthropological account of sericulture in Karnataka, see Charsley (1982), and in India, Acharya.
14. One of the Maratha households which owned this tiny plot subsequently sold it to an Okkaliga landowner.
15. The phrase means "the land lying immediately next to the village settlement".
16. The term refers to someone who is a well-to-do farmer or householder.
17. For evidence of persistence of concealed tenancy and sharecropping elsewhere in Karnataka, see Damle (1989) and Aziz and Krishna (forthcoming).
18. For a detailed account of types of agricultural labour arrangements, see Chapter V.

2

Agrarian Relations Prior to 1950s

An attempt is made in this chapter to present a brief history of the village and its agrarian relations in the past so as to trace and analyse the change and continuity in them during the succeeding years. The choice of dividing line as 'prior to 1950s' is because until then the village may be viewed as reaching a stage of formation and elaboration of agrarian relations, having come into existence during 1850s. Furthermore, since 1950s the village witnessed certain changes that were crucial for us to mark it as a watershed in its economic and social history. The history as traced here is a reconstruction based on information gathered from informants both in the village and the region, as well as on perusal of land records.

Historical Background of Rajapura

A passer-by near the Sidda Betta forest[1] on the southern border of the village is shown a place which the local people claim to have been the original settlement site of the people of Rajapura about a hundred years ago. This site is referred to as Sathe Katte where evidence of an earlier inhabitation of people can be seen. Much of the area is now classified as forest land. It is said that people migrated to Sathe Katte from a nearby village Neralawadi, lying south of the forest. Although the exact reason for their migration is not known, it is possible to assume that they obtained a grant of land as *raktha kodige inam*[2] in the neighbourhood of Sathe Katte. These lands are today part of the village Rajapura.

One of the neglected aspects of village studies in India is tracing the origin of the villages, particularly of those with a recent origin.[3] Awareness of such a history helps one for a better understanding of the regional variation in agrarian and social structure. A cursory perusal of the successive Census reports in India reveals that almost every district in each state contains villages that were earlier uninhabited or have now become one. In Madabal hobli of Magadi taluk, of which revenue administrative unit Rajapura is one village, there are many such instances of depopulation of inhabited villages or transformation of uninhabited villages into inhabited villages. Such demographic changes have occurred mainly in the former inam villages which had a privileged tenurial rights and obligations to the state. Little attention has been paid to such changes by students of rural India. Many such uninhabited villages today have human presence, and the lands in such villages are now cultivated by people living in either distant or neighbouring villages. One may find little information in the Census reports regarding these rural dynamics. It is possible that the people left the villages because of frequent famines or epidemics or also because of the threat to lives and property from the wild animals and thieves.

Rothermond (1977) in an analysis based on the reports of the district officers during the British rule finds evidences of desertion or appearance of new villages in the country. In various parts of the country the district officers noted that what is now wasteland had been cultivated in the past, and remnants of buildings and tanks etc., existed in such areas. In Karnataka, particularly in the old Mysore region, it is possible for us to find a parallel of this nature. During the reign of Tipu Sultan there are evidences of desertion of villages by cultivators owing to the hardship they faced at the hands of the king's representatives, and the political instability during the wars with the British. When the Sultan raised the tax to be collected from the farmers in 1786 to meet the expenses of the war in 1780-4, the cultivators who were unable to bear the increased taxes, fled the villages. When this was noticed, the Sultan introduced various measures to attract the cultivators to return to the villages, and to prevent the officials from misusing their positions he fined the concerned officials if cultivators fled the village or gaveup cultivation of land (Gopal 1971).

The region surrounding Rajapura too has witnessed many petty and major wars of the rulers against the British. Notable among them was the one fought by Tipu against the British at Savandurg, barely three kilometres as the crow flies north-east from Rajapura. Buchanan (1807) notes that the area surrounding Savandurg to have been under thick forest following the desertion of people after the fall of

Savandurg to British forces. Today, Savandurg is a state forest. Villagers now point out to areas around Rajapura by names that had been formerly a village or a hamlet of people living there. Some such places have found their way into the list of villages in Census reports, while most others have not. For instance, places like Sathe Katte, are part of one or the other village around Rajapura, the settlement having been shifted to the present village site.

Evidences from Buchanan's report and also from the writings of Gopal regarding Tipu's rule suggest that there had been some demographic changes in the region surrounding Savandurg. It is, therefore, possible that the original inhabitants of Sathe Katte moved into the present site of Rajapura as a result of political unsettlement and also because of a land grant made to a family of this area. This grant was made to a family near Rajapura, and this land is now under the revenue jurisdiction of Rajapura. The descendants of this *inam* recipients continue to live in the village, who also possess some arms like sword and spears believed to have been used by their ancestors. Gifting of this land may have been an incentive for original migrants to more either from Neralawadi or from Sathe Katte. Moreover, it is said that the people of Sathe Katte faced constant threat to lives and property by the *pindaris, kandars*, and *panjukallas*, all of whom were noted bands of thieves and robbers in the region in the past. The changes in the political administration in terms of the transfer of power from Tipu to British, and then to the former princes of Mysore may have resulted in law and order problems in the region (see, Hetne 1978).

Furthermore, the First Survey and Settlement Report for Madabal Hobli (1879) notes that the region of Magadi, particularly Madabal Hobli to be famous for the 'fever '(Malaria ?) and the 'wilds of the jungle'. The cattle of the region even now face constant threat from the wild creatures living in the forest of Sidda Betta and Savandurg. Coupled with the threat to lives of people and cattle, and being nearer to the site of *inam* land the people living in Sathe Katte migrated to Rajapura. When compared to the topography of Sathe Katte, Rajapura is plain and had the safety in terms of having other villages close-by in the neighbouring area.

It is difficult to determine the exact date of origin of Rajapura settlement. The first ever census taken in 1871 lists Rajapura as a village while making no mention of Sathe Katte. The village must have, therefore, come into being earlier than 1871. According to 1871

Census, the village had a population of 236 persons. The Survey and Settlement Report of 1879 mentions that the village had only 104 persons that year. The decline in population in a short period of eight years may be accounted for the "fever" and the "wilds of the jungle" (Department of Revenue, Survey and Settlement Report 1879). The genealogies of the people living now in the village and the analysis of land records suggest that the original group which settled in the village consisted of about six to seven households who had migrated earlier from Neralawadi, Kottagana Halli and Lakshmipura, and much later from Madabal. All these villages are within a radius of about ten kilometres from Rajapura. Members of these households were all kinsmen or agnates even when they were living in their native villages. Having migrated to Sathe Katte and later to Rajapura, they invited others to join in developing the virgin land and settle down as cultivators. Gradually, the number of the inhabitants increased when the relatives and others began to settle in the village. While most of the founder settlers were all from Okkaliga caste, there were several households belonging to artisan and service castes who accompanied them in founding the new village. The Holeyas and Madigas (members of the former untouchable castes, who were traditionally the agricultural labour castes in the region) reported that their ancestors migrated from villages like Yelige Halli, Neralawadi, Chakrabhavi, etc. It is said that the lands situated in the outskirts of the village were all under forest, and that their services were hired for clearing the land for cultivation.

Around this time, that is 1870s, one vast area of land attracted the inhabitants of Rajapura. This land, forming one *inam* village, Ramapalya, belonged to a Brahmin living in Magadi town. It is said that even long after Rajapura had developed into a well settled village, the land owned by the Brahmin was under thick forest, and is even now referred to as a forest. Since the agrarian history of Ramapalya has played a crucial role in the evolution of agrarian relations in Rajapura over the decades, it is useful here to briefly trace the antecedents of Ramapalya. A few Korama caste households, basket makers by profession, and Ilugas, a tribe, were inhabiting this forest. The Koramas had also cleared some part of the forest and were cultivating the land unauthorisedly. This patch of land is even now referred to as Koramara Palya, meaning the settlement of Koramas, and is part of Ramapalya. One of the Koramas claimed that his parents migrated from Koramara Palya to Rajapura following their eviction by the landowner. A few Iluga households now living in the nearby village of Ajjarahalli, 5 kilometres from Rajapura, too informed that their parents were living on the lands of Ramapalya before the tenants of Rajapura together with the land owner from Magadi forced them to leave the place.

A descendent of the Brahmin landowner, now living in Bangalore city, informed that his ancestors received the grant from the rulers of Mysore in appreciation of a musical performance of a member of their family. Available evidence, however, does not confirm this claim. A list of villages published by the then Government of Mysore[5] lists this village as a *jodi inam*, that is, a village with a light assessment of revenue than normal (Rao 1929). A much earlier report, the Survey and Settlement Report of 1879, lists Ramapalya as a village of *Khayam Gutta* assessment. *Khayam Gutta* villages are those for which revenue assessment has to be paid once and for all, thereby the farmer or owner enjoys rights over the land for a specified period. The revenue papers pertaining to Ramapalya refer to it as a *Khayam Gutta* village. Whether it was a *jodi* or *Khayam Gutta* village, it was an *inam* village, i.e., it had privileged tenurial rights. This is evident from the fact that in 1956 it was affected by the *Inam* Abolition Act of the then Mysore State.

A few households are reported to have lived in Ramapalya during the Census of 1941 and 1951, but it is apparent that people lived there temporarily. The 1951 Census informs that the residents of this village were all 'tenants cultivating the lands of others', none being 'resident cultivators of own land'. From my study I learn that all along Ramapalya lands were cultivated by tenants and sharecroppers of Rajapura. A few years prior to the enactment of legislation abolishing *inamdari* holdings did a few tenants from the neighbouring village of Kuppe began to cultivate the land. This was following the eviction of some tenants from Rajapura for fear of their claiming permanent rights over the land. Today, more or less, all the land in this village is owned and controlled by farmers of Rajapura.[6]

At the turn of the present century, the houses in Rajapura were located around the shrine of the presiding deity, Maramma. Based on the information given by the local people, particularly the descendants of the founders of the village, it is possible to deduce that there were about 15-20 households living in the village at that time. The available Census and other data regarding the population and number of houses till 1971 and of 1978-79 are presented in Table 2.1. Between 1871 and

1879, a span of eight years, there was decline in population by more than half. It is possible that this decline was due to under reporting, or a high rate of mortality due to epidemics and wild animals. It is also possible that many of the people who had originally settled in the new village returned to their native villages either because of the improved conditions in their native villages or due to inadequate facilities of social, ritual and economic requirements in the newly founded village.

TABLE 2.1: Growth of Population in Rajapura

Year	House* House- household	Population	Average per household
1871	46	236	5.1
1879	31	104	3.3
1941	NA	487	NA
1951	112	657	5.9
1961	165	901	5.5
1971	143	796	5.6
1978-79	183	995	5.4

Sources: 1871; 1951; and 1971 Census reports; and 1879 Survey and Settlement Report for Magadi Taluk.

Note: **Houses* in 1871 till 1951 refers to households, whereas the Survey and Settlement Report defines every 'door' in the village as a 'house', which may be misleading.

In the absence of any documented data all these inferences remain conjectures. However, there was steady increase in the population of the village in the subsequent period till 1961, after which again there is a decline in the population.[7] During 1964 and 1968, the region faced one of the worst droughts which compelled many people to leave the village in search of food and occupation. Many persons living now in the village reported of their having gone to places as far as Kodagu and Chikkamagalur districts where they had attached themselves to the coffee estates for years together. Some Madiga labourers had gone to live in Bangalore city and Mandya region in search of employment in the industrial establishments, agriculture, construction services, etc., to return when conditions improved.

TABLE 2.2: Pattern of Landownership (1930s)

Caste	Land Ownership by							
	Residents			Absentees			Total	
	No.	Area	Average area per household	No.	Area	Average area per household	No.	Area
Brahman	—	—	—	5	48.02	9.60	5	48.02
Lingayat	1	4.10	4.10	6	91.14	15.19	7	95.24
Okkaliga	28	286.14	10.22	9	58.97	6.55	37	345.11
Maratha	—	—	—	3	84.20	24.07	3	84.20
Vaishya	—	—	—	3	17.65	5.88	3	17.65
Lambani	1	4.00	4.00	—	—	—	1	4.00
Holeya	9	40.00	4.44	—	—	—	9	40.00
Madiga	14	51.11	3.65	—	—	—	14	51.11
Muslim	—	—	—	5	42.53	8.51	5	42.53
Tank, village site, waste, etc.		184.56						184.56
Inam: Temple		34.28						34.28
Total*	53	604.19	11.40	31	342.51	10.70	84	946.70
Ramapalya Lands				1	159.00	159.00	1	159.00
Grand Total	53	604.19	11.40	32	501.51	15.67	85	1105.70

Note: * The 'total' refers to the land owned by individuals living in the village and for the land owned by them within the village.
Source: Index of Land kept in the Taluk Revenue Office for Rajapura and Ramapalya.

Landownership Pattern :

The First Survey and Settlement Report also notes that land in this region have been mainly under forest growth. The settlement of people in Rajapura at the turn of the century also initiated a process of clearing the jungle and land reclamation on both the lands held by individuals and those of the forest. The Settlement also streamlined the rights over land. Vast areas of land were classified as cultivable and sold to people through auction. A review of land records suggests that much of the land was bought by absentee landowners, i.e., those who were residents elsewhere (Table 2.2). This has implication for the mode of cultivation: that is through tenants and sharecroppers.

During 1930, when the Second Survey and Settlement was carried out, 36.2 per cent of the cultivable land was owned by absentee landlords, most of whom had non-agricultural occupations. Out of the 53 resident landowners a majority were Okkaligas, who constituted nearly 53 per cent of the resident landowners, and owned nearly half the land owned by resident landowners. Okkaligas, whether resident or absentee landowners, as a group owned 47.0 per cent of all the land in Rajapura. The Holeya and Madiga castes also held land most of which were granted to them by the Government under the Grow More Food Scheme. The Vaishyas were all from Magadi town engaged in trade, and had come to own land in Rajapura initially through moneylending. The Muslims were those frequenting the village to fell trees and make charcoal. They too indulged in moneylending d later acquired the land.

Land had entered the market as early as 1920. A major share of transaction occurred through the government granting occupancy rights over the lands that were cultivated unauthorisedly for a few years, and ownership rights to applicants for government land (*Darkast*). Apart from such buying there used to be auctioning of land. Usually a select few would learn in advance of such auctions or sales of land by the government. However, not all farmers were in a position to take part in bidding for land when auctions were held, or when sale of land occurred. The former *Patel*, *Shanubhog* and the person said to be the founder of the village were in a better position in this respect. Usually they were aware of a forthcoming sale or auction of land, and were thus able to buy a good deal of land rich in irrigation and proximity to the village. The tenants of *shanubhog* and many others working in the bureaucracy were also benefited by such land transfers.

A few aged farmers today recall that these officials had persuaded them to buy land, for, occasionally there had been no buyers. There are also instances of the *shanubhog* lending the money to farmers for buying land.

The absentee landowners usually had a non-agricultural income, which enabled them to invest in landed property. They consisted of a few officials of forest, and revenue departments, merchants, and kinsmen of the *shanubhog*. Not only did these men have access to information regarding land transfers, but also sources of cash income which helped them in buying land. A second manner in which land transfers took place during the period was due to the land grants made to the members of depressed castes and classes. Under the Grow More Food Scheme many landless and weaker sections of the society were given waste and grazing land situated in the villages. On an application made by the former untouchables and other poor people, which in turn had to be recommended and certified by the village officials, the government granted such lands. At the time of sanctioning the rights of occupation and ownership, the grantee had to pay a nominal fee towards the measurement of land, marking the boundary, and registration. It was binding on the part of the grantee not to alienate such land for six years[8], except through inheritance and mortgaging for loans in cooperative credit institutions. While the granting of these lands in itself is a manner of land transfer, it also initiated a further process of large scale transfers — a process which did not remain restricted to the period under discussion. The recipients of such land grants usually were members of the former untouchable castes, who held no other resources at that time except their labour. They had been placed in a difficult situation in raising the necessary funds for the payment of tax, land price and registration expenses. Even if they were able to raise the sum without having to incur a fresh loan, they were left with no capital for investing in buying the necessary seeds, tools and implements as well as cattle. Although the quality of land granted during this period was much better than those granted during 1970s, the grantees had to invest money in clearing the land suitably for cultivation. Most of such land grantees were bonded labourers, and had no free time at their disposal to devote full time on their newly acquired land. As a result of all these, the grantees had to depend heavily upon the moneylenders either within the village or elsewhere. Thus the newly granted land always ended up being pledged for such loans. Indeed many merchants belonging to the

Vaishya caste had advanced money to the farmers of Rajapura with land being pledged. The labouring Holeyas and Madigas had another source in the moneylending Muslim traders engaged in felling trees in the forest, for whom many from these castes worked regularly.

Thus, the wealthy moneylending landowners, merchants and visiting traders advanced money to the farmers and began to gain control and ownership of land. Since the grants were made on the condition that it shall not be alienated before six years, much of such transactions were informal and oral. However, when the minimum period expired the transaction was recorded in the books maintained by the village accountant.

It is important to mention here that the manner of land transfers for most Holeya, Madiga, and Lambani landowners began as a loan transaction. Usually, while borrowing money their land is pledged, and the terms of the transaction is such that the amount borrowed, with interest, is repaid within a specified period. Failing a timely repayment, the land is deemed to be sold. In revenue terminology it is referred to as 'condition sale' (locally, *nambike kraya*). However, the borrowers were usually unable to repay on time. Instead, there was a tendency to incur fresh loans. There is one landowner in the village today, an Okkaliga, who owns mainly the lands alienated from the land grantees: out of his 24 odd acres, nearly 20 acres were formerly owned by the land grantees of Holeya and Madiga castes. One Madiga grantee had an interesting tale to narrate which points out the process and extent to which his caste men were habituated to borrowing money from moneylenders.

A hunter before setting a trap for birds under a tree regularly fed them on a particular seed for a few days. After a while the birds became addicted to these seeds. One day the hunter skips feeding the birds, and the next day feeds them the seeds— but only after setting the trap. The birds that would have starved overnight for want of seeds flock to pick the seeds and get caught in the trap.

In this story, the former untouchable raconteur compared the hunters to the Okkaliga moneylenders, birds to the former untouchables who were granted the lands, and seeds to loans of money and grains.

Some Features of Agrarian Relations

The dominant feature of agrarian relations during the period prior to

1950s was that nearly half the land in the village was under the control and ownership of outsiders. Many of them were with non-agricultural occupations, and were living in Magadi, Ramanagaram and Bangalore. They were either traders or employees in government departments. Some of them had access to information regarding the availability of land for sale from time to time and were also in a position to pay for the land in cash unlike many farmers in Rajapura. The tenants of the hereditary village officials and their close relatives were also able to buy a considerable extent of land. One finds that these people were thus privileged to buy the prestigious *ooru pakkada jameenu* and those which contained sufficient irrigation potential. Indeed an extent of 18 acres adjacent to the village site having the best underground water resources in the village was owned by a Brahman landowner related to the *Shanubhog*. During the 1930s, when the resurvey of land was carried out, land lying adjacent to the village site was owned mainly by the *patel*, *shanubhog*, the founder of the village, recipients of the *raktha kodige inam*, the village servants and one Lingayat merchant. As a result of large extent of land being held by absentee landowners, cultivation was largely through tenants and sharecroppers. In sharecropping system (*varam*), the sharecropper had to share the produce equally with the landowner. He was also required to give a specified quantity of fodder. The cost of cultivation was to be borne by him, in addition to the cost of transporting the landowner's share of produce. Under the tenancy system (*guttige*) the tenant paid a fixed rent either in kind or cash, or both to the landowner. In contrast to the sharecropper, he had to undertake to pay the rent whether or not the crops were good. Although the tenant is not bound to give a portion of the produce of each crop while paying the rent in kind, he usually supplied vegetables, fruits, groundnuts, and the like to landowner. Such friendly acts usually helped the tenant in maintaining a cordial relation with the landowner and ensured a continued tenant-landowner relationship. It was not, however, uncommon for a landowner to have been displeased with a tenant if the latter failed to give in kind and in plenty in addition to the rent as agreed upon.

It was a common practice for an absentee landowner of this period to channel his relations with tenants and sharecroppers through a representative from among them. Such a representative, called himself a manager, functioned as a go-between the other tenants and landlord. He supervised the collection of rents and the share of the landlord and acted as a supervisor of the property. As the village population grew, specially because more and more families were settling in it as there was plenty of land available for tenancy or sharecropping, the role of manager became quiet important. Such representatives were in a position to procure land for the aspiring tenant-cultivators and also causing the eviction of earlier tenants, at their will.

The growth of the population had a few consequences for the village. First, there was a growing competition for land for tenancy and sharecropping. Secondly, in turn, the rent that was being paid also began to reach higher levels. Third, gradually there was a decline in the fixed rent under tenancy, and a rise in the sharecropping system of cultivation. Fourth, as a result of the increased competition, there was greater insecurity of tenants. There are innumerable instances of rivalry that began during this period between the tenants which has persisted even today.

In most cases, the tenants and sharecroppers of the period were owning some land of their own. This was particularly so with Okkaligas. However, the immigrants who had migrated on an earlier agreement or understanding of gaining an opportunity to sharecrop with the help of resident landowning relatives, were to be provided with land for cultivation either for ownership or on tenancy. Hence, each tenant or an aspiring tenant was vying with each other by offering higher rent or crop share. Thus, from one-third the share of the landlord was increased to equal share.

The landowners residing outside the village received the rent in kind or cash, which was delivered at their door-step, transported in bullock carts. If the tenant failed to deliver the cropshare or rent on time, the representatives were empowered to demand it from the tenant. Ultimately the landlord himself visited the village to collect the dues. However, it was a common practice for the landlords to visit the village during the harvesting of crops to ensure that the crops were equally divided as shares, and that the quality of grain was same as the one which the farmer retained with him. The landlord also used to visit the farms to collect a share of other crops grown on the land, particularly *avarekai*, groundnuts, and jack fruits and mangoes.

Although Okkaligas were the ones with maximum land under their control and ownership, they also held much of the land under tenancy cultivation. As noted above, since the 1920s the village was experiencing a large scale immigration. Such a process had been taking place mainly because there was plenty of land available for those who could afford to buy or for those who wanted to be tenants or sharecroppers. As members of the dominant caste the Okkaligas were able to sponsor many of their relatives immigrating to the village. They were also able to procure tenancy occupation to such immigrant kinsmen if land was not immediately available for buying. However, the caste dominance of Okkaligas over all the others was now and then faced with challenge from the lower castes who also attempted to gain a relative economic independence by virtue of tenancy rights for lands. The Okkaligas, on their part, were able to curtail such an attempt at independence on the part of the lower castes. The case of a Madiga tenant may be cited here. He was cultivating nearly ten acres of land as a tenant of a few absentee landowners. He had sub-let a portion of this to the fellow caste members, thus gaining a higher economic and social status among his people. Unlike many of his caste men, he was not dependent upon the Okkaligas of the village for his social and economic needs. On the contrary, he had been lending money and grain to the members of his caste, and had even employed a few bonded labourers. The relative independence and his verbal expression of it was resented by many Okkaliga leaders. It was not the purpose of the absentee landowners to encourage the economic and social independence of the Madiga tenant. Absentee landowners cannot for too long be alienated from the local leadership, and therefore, Okkaligas were ultimately successful in getting the Madiga tenant evicted from the land. In fact, part of this land was bought by the local Okkaligas. Not contented with their success, the local leaders further prevented him from obtaining a land grant on the pretext that the allotted land was an important passage for the cattle to go to the forest for grazing. The same plot of land was later allotted to an immigrant Okkaliga. A few years later the Madiga tenant was excommunicated from the village for what was alleged to be an act of disrespect towards the village elders and for violating the caste taboo on coming in contact with the village drinking water well. In due course, he lost all his wealth, and his children had to enter bonded service with a few wealthy farmers.

Indeed the factionalism that exists in the village may be traced to, among other factors, a few such incidents involving rivalry and competition for use, control and ownership of land. So much so that rival groups had been sponsoring their kinsmen to migrate so as to numerically challenge the dominance of the village elders.

Agricultural Labour

Agricultural labourers were drawn from all castes. There were also many Okkaliga households, mainly the immigrants, whose members had been working as agricultural labourers. Some were working even as bonded labourers. Such Okkaliga labourers were mainly those who were at that time yet unsuccessful in gaining tenancy or be able to buy land. However, the majority of labourers were Oddas, Lambanis, Holeyas and Madigas. The Lambanis were provided with a separate colony (*thandya*) in the outskirts of the village during 1930s. They were also provided with some land for them to be able to settle and take to cultivation. Like the Holeyas and Madigas, they too alienated their land to Okkaligas and to the Muslim traders. The Okkaligas were also eager to own land for themselves or for their sponsored immigrants. Hence, they were willing to advance loans to the land grantees among the labourers either on interest, mortgage of land, or attaching the labourers. The granting of land to many landless labourers in those days did not improve their living conditions. On the contrary, it encouraged indebtedness among the neo-landowning class of lower castes. Many of those who have lost their lands informed that they were not usually free to attend to their land but committed to work on the land of the wealthier landowners either because of a traditional obligation or having to repay a loan made earlier during the year.

One of the interesting aspects of looking into the features of landowner-labour relations in the past, as recalled by the people now, is a perfect, harmonious, benevolent — in short, glorious — relationship between the two. This, however, does not mean that there was no disharmony or conflict. Let me first present the 'glorious' picture as reconstructed through people's narration. A labourer was expected to start working at sun-rise. He would work with a commitment as though the land belonged to him, without having to be supervised constantly. Often the labourers took the initiative in the work by reminding the landowners of their responsibilities in accordance with the demands of the season. The landowner did not have to repeatedly call them to report for work. The labourer had a long-standing association with the landowner. So much so, he had a complete knowledge of the number of cows, bullocks, ploughs and other implements owned by the landowner. Tools and implements needing repairs were taken by the labourers to the iron-smith by him without the landowner having to specifically instruct him. There was a sense of involvement on the part of the labourer not only with the landowner and his household, but also towards the village's well being.

In keeping with this, the landowners did not hesitate in advancing loans to the labourers, or helping them in times of their need. Grains were lent when the labourer was out of work, or was asked to eat the left-overs in the employers' house. The size of ragi balls (the staple food) was "much bigger, and tastier". The quality of the grain given as wage was much better, and the measures reliable. The generosity on the part of the landowners was such that the labourer was given building materials while constructing a house, or given money to meet the medical expenses or expenses of a marriage, birth, or death in the family.

It hardly needs to be stated that this 'glorious' picture is completely true or false. It is true that the farmers have helped their labourers, but with certain interests to be fulfilled. Recall the tale cited earlier, in which the 'trap' has implications for the 'generosity' of the landowners. Thus, the harmony in the past was a coercive and not a consensual one. Furthermore, whatever the motive behind the generosity of the landowners, the same kind of acts of generosity are not totally absent in the present day agrarian relations. I shall return to a discussion of this in a different chapter.

Jajmani Relations

It is difficult to pin-point the time around which the *jajmani* system developed fully in Rajapura.[9] A few service specialists had migrated along with Okkaliga landowners when the village was founded. All the musicians had come from Neralawadi, which was the village from where the founders of Rajapura had come. Further, that some of these relations had taken strong roots into the village's agrarian social structure is evident by the fact that the Second Survey and Settlement had recorded in the *Index of Land* the *inam* land for priests, village accountant, and the *thoti* (village servant).

Thus, by the year 1930, there had been service specialists like a few priests, an astrologer, messenger, barber, washermen, potter, musicians, and an iron-smith. With the exception of the astrologer and barber, all other specialists were residents of Rajapura. The local village panchayat looked after the smooth functioning of the system involving different specialists from different castes. The Panchayat also took decisions regarding the plea made by new and aspirant service specialists to find a place in the network of *jajmani* relations in the village, or in letting them to serve in other villages. Each caste had its own council which also gave directives to the members with regard to their occupations and their role in the *jajmani* system.

Disputes were not uncommon both with regard to the right to be served and to serve, as well as regarding the payment of grain. The respective caste council and village council jointly resolved such disputes occurring within a caste. However, it was not uncommon for the village panchayat to have a decisive say in matters of intra-caste dispute. There has not been an instance of the caste council taking a decision which went against the wishes of the local leadership, although the actions of the latter had more than once violated the norms of exclusive rights of a specialist in his chosen field of specialisation. Competition between specialists of a caste occupation was permitted by the village council even though the caste council had opposed it, as in the case of the three iron-smiths, and two washermen in the network of *jajmani* relations.

Notes

1. See Chapter 1, for a reference to this forest and its location.
2. *Rakta Kodige* is a grant of land to the survivors of a war hero who shed blood or gave his life braving a war for the ruler.
3. There are some exceptions. See, e.g., Chauhan (1967), Kessinger (1974). See also, Shah (1974) for a review of studies in social history.
4. The 1871 Census report states that in a number of villages there was widespread under-reporting of people living in its households for the fear of compulsory recruitment in the armed forces by the British. The same may have occurred again in 1879 during the Survey and Settlement.
3. *List of Villages in the Mysore State*, Government Press, Bangalore 1905.
6. See Chapter 3 for the description of the process of this transfer of lands to the farmers of Rajapura.
7. Informants in the village, mainly the leaders, recollect that the population figures for 1961 Census were deliberately inflated in order to show a higher population, so that the village could gain the status of being the headquarters of the village group panchayat. This ultimately went to the neighbouring hamlet Doddi, where the leaders had manipulated the population figures even higher than that of Rajapura. Doddi is only a hamlet of the village Kuppe, and does not even appear in the revenue map as a village or hamlet.
8. Subsequently, it was raised to ten and later to fifteen years.
9. The scope for the discussion of *jajmani* relations here is limited only to refer to its emergence in the village. For a detailed discussion of this aspect see, Karanth (1987a).

3

The Land Reforms Era

The 1950s heralded a new phase in the agrarian history of Rajapura, which, however, was in keeping with the political developments in the state and the country. The state, then known as Mysore, began to formulate and implement measures of tenurial reforms, particularly the *inam* land tenure and tenancy reforms. Prior to 1947 too, the state had taken some steps in this direction. These measures, however, lacked a commitment to the regulation of relationship between tenants and the landowners, and were mainly oriented towards the *inam* villages rather than the non-inam villages.[1] These earlier measures were mainly advantageous to the *inamdars* and other landowners rather than to the tenant cultivators. Some measures like the Inam Rules, or the Land Revenue Code of 1888, with subsequent amendments, referred mainly to the recovery of dues from the tenants and provided little security to the tenants. The interests of the tenants and sharecroppers were by and large ignored.

Evolution of Land Reform Legislation in Karnataka

Land reform measures after 1947 had an impact on the tenurial conditions in rural areas, mainly due to the time gap between the announcing of land reform measures and its implementation. Indeed the lengthy period prior to the implementation brought about much changes in the landownership pattern defeating the purpose of reforms. Initially, the state government appointed a committee — The Committee for the Revision of Land Revenue System in Mysore — in 1948 to examine the feasibility of simplifying the land tenure system by abolishing inam tenures in the state. The Committee in its report recommended the abolition of *inam* tenure among other things. Even before any steps were taken to implement the recommendations, there were large scale eviction of tenants and transfers of land by the owners.

There, in order to prevent eviction of tenants and to reduce the rent for land, the Mysore Alienated Villages (Protection of Tenants and Miscellaneous Provision) (Emergency) Act, 1949, was passed. The Act aimed at protecting the tenants of *inam* land who were cultivating successively for the previous five years. It was only in 1954 that an Act was passed abolishing the *inam* tenure. This Act known as the Mysroe (Personal and Miscellaneous) Inam Abolition Act, 1954, aimed at abolishing all major inam holdings excluding the village service *inams* held by the hereditary village officials.[2] The tenants of such lands were permitted to buy the land by paying the land revenue and premium for ownership that was equal to 20 times the land revenue. Later, in the year 1955, a further enactment was made to cover the *inam* land of religious and charitable institutions. Under this Act, the tenants of temples and charitable institutions' land were given protection similar to that under the 1954 Act. Persons rendering services in the religious institutions (such as the hereditary priests) were exempted from the Act, and were allowed to enjoy the land. Tenants of *inam* land were treated as equal to the tenants of revenue land. Provisions were also made for hereditary occupancy rights for such lands provided that the rent was paid regularly, that they maintained and cultivated the land without leaving it fallow, and did not alienate the land for non-agricultural purposes.

With respect to revenue lands, a committee[3] was appointed in 1957, with a view to examine the feasibility of preparing a common tenancy law in the state of Mysore following the reorganisation in 1956.[4] The Committee had to examine and suggest uniform measures for fixity of rent, security of tenure, right of resumption of land by the landowners for personal cultivation, right to purchase land by the tenants, payment of compensation to the landowners, ceiling on holdings, prohibition of landownership as a source of income by the absentee landowners, and prevention of land alienation for non-agricultural purposes. The recommendations of the committee, described as the "fountain head" (Rajan 1979), formed the basis of land reforms legislation in Karnataka during the years to come. However, the enactment basing on the recommendations of the Committee could not be done till 1961, when the Mysore Land Reforms Act, 1961 was passed, and its implementation till 1965.

The 1961 Act contained many liberal clauses under which certain classes of landowners were not only permitted to lease their land on tenancy but also resume land for personal cultivation. For example, the persons serving in the Defence forces and in the Merchant Navy, as well as minors, unmarried women, widows, persons suffering from physical and mental disabilities, and small land holders were all permitted to lease their land. With certain conditions, resumption of land for personal cultivation was permitted. The Munsif Tribunals were constituted for the purposes of dealing with the applications for resumption of land, but these tribunals were not equipped with adequate personnel. One of the ministers involved in the formulation of the 1961 Act later wrote:

> ....as the requisite number of tribunals had not been appointed and as the Government did not issue certain notifications which were essential for the implementation of the Act, the progress of implementation was very tardy.... (Further, taking advantages of the delay in the implementation of the Act) powerful landlords evicted their tenants, by force in some places. In some others, landlords, in connivance with unscrupulous village accountants, got their names entered in the revenue records to create false evidence of their personal cultivation (Manjappa 1984) (Parentheses added).

As a result of the loopholes in the 1961 Act and the procedural delay, the Act failed to confer occupancy and ownership rights on the tenants. The Act of 1961 was amended in 1974, aiming at a fundamental change in the land relations. The new Act withdrew several of the concessions given earlier to the minors, widows, etc. Exemption was applicable only to the soldiers and the seamen. All leased land stood vested with the government. The right to resume land for personal cultivation was also withdrawn. With further amendments, the tenants were subsequently permitted to submit their applications for confirmation of occupancy rights with the land tribunals, until June 30, 1979. The new land tribunals, replacing the earlier Munsif Tribunals, consisted of one ex-officio member, represented by the Assistant Commissioner as its chairman, and four non-official members including a member of either the Scheduled Caste or Scheduled Tribe. The tribunal members were asked to examine the application of the tenants and pass their orders in accordance with the Land Reforms Act, 1974, by taking into account the evidences through persons in the village and records. The decisions of the tribunals were final; the aggrieved could seek redressal only through a writ petition to the High Court in the State. The High Court, in turn may pass orders to the tribunal directing it to reconsider the procedure of its arriving at the judgement.[5]

Landowners and Tenant Relations in Rajapura

By 1950, Rajapura had both *inam* and *ryotwari* tenure of land in practice. The three prevalent *inam* tenures can be classified under the following categories: (i) lands which were under individual control (personal *inam*); (ii) under the control of institutions such as temples, (*devadaya*); and (iii) service *inams* held by hereditary accountant, and messenger (see Table 3.1).

Besides the three temples of Rajapura, two other temples in Magadi also owned about 15 acres of land. While the priests of the local temples cultivated its land, the other tenants cultivated the land belonging to Magadi temples, accountant, and the *jodi inam* of Ramapalya. Rights to cultivate the land of the two temples were auctioned once in five years, a system known as *panch sal guittige*.

TABLE 3.1: Inam Tenure in Rajapura (1950)

No.	Inam Types		Area in Acres
1.	Temple (*Devadaya*) *Inam*		44.43
2.	Service *Inam*		
	(i) Village Accountant	3.53	
	(ii) Village Servant	4.95	8.48
3.	Personal *Inam*		
	(i) *Raktha Kodige Inam*	26.28	
	(ii) *Ramapalya jodi inam*	159.00	185.28
	Total		238.19

Source: Index of Land

The *ryotwari* holdings of the resident farmers were mainly under personal cultivation. As noted in Chapter 2, the bulk of the land was owned by the absentee landlords, who leased their land to the tenants through their representatives living in the village. Lease contracts were usually for a period of one year, renewable at the end of the term. Continuation of the same tenants for the same plot of land depended upon tenants's personal relations not only with the landowners but also with the representatives, if any. It was also observed that with the availability of land in plenty, there had been a steady immigration to the village which, in turn, had contributed to competition for tenancy. This phenomenon was specially applicable to the lands of Ramapalya, where there used to be frequent eviction of tenants owing to the competition among farmers.

The absentee landowners were, by and large, dependent upon the local leaders, particularly of Okkaliga caste. Whenever there were problems of irregular payment of rents or crop-share, or loans were overdue their intervention was sought. There were also instances of the village leaders bringing pressure upon the absentee landowners to prevent displacement or eviction of tenants whom the village leadership favoured. Often, the leaders themselves had recommended some tenants to the landowners, and thus had both tenants and landowners under obligation. This position of influence enjoyed by the leaders assumed great importance in the years following the enactment of land reforms laws.

Impact of Early Legislations: A Few Case Studies

The early '50s saw the enactment of some important legislations pertaining to *inam* land tenures. These Acts and their subsequent amendments brought about certain changes in the agrarian system of Rajapura. It was earlier pointed out that nearly 500 acres out of 886 acres of land in both Rajapura and Ramapalya were held by the absentee landowners, and that almost all such lands were cultivated by the tenants and sharecroppers. These Acts brought some awareness among farmers and absentee landowners of the impending changes within the agrarian system. Many shrewd absentee landowners started disposing off their land well before the axe fell. Many scholars have noted this phenomenon. For example, Ladejinsky observed[6]:

> Reform was in the air before the enactment of the laws; and this, combined with the long delays in passing the law, made it possible for the landlords to reduce the number of claimants for rights in the land by evicting or shifting them to the status of farmhands. (1977:378)

Given their economic conditions, not all tenants were in a

position to buy lands that were being sold by the landlords. Added to this, there were many aspirants for these lands for other reasons. Important among them were: those who had owned land adjacent to the land being sold were also keen on buying it; because such lands were near the village settlement; and competition among farmers to buy land for their kinsmen who had migrated in the hope of buying land.

As a result, land price began to soar, thus rendering it beyond the reach of most tenants. In some cases, tenants themselves were owners of small holdings and were not interested in buying the land even if it were offered to them for purchase. However, the tenant was not always offered the land for buying. When a tenant was interested in buying, he had not only to seek the goodwill of the landowners, but also of the representatives and local leaders. The village leaders played an important role in all such transactions. An analysis of a few cases reveals the role of local leaders as not being consistent: while in some they sided with the landowners in some others they sided with the tenants and other local buyers. In some of the tenanted lands, the leaders themselves were interested in buying them even though they had not been tenants which meant depriving the tenant the land.

With regard to the tenancy arrangements for the land owned by the *resident* landowners, as noted earlier, they were, by and large, unwritten, informal and on an *ad-hoc* basis. Consequently, none of these faced a threat from the land reform laws. Whenever a formal arrangement existed, by way of a written lease contract, etc., the landowners resumed their land for 'personal' cultivation by evicting the tenants. This was not a difficult task as both the landowners and the tenants were from within the village and pressure from caste, kinship and political dimensions were applied. Further, having let out a portion of land, the local landowners usually cultivated the remaining land by themselves. This enabled them to get their own names entered in the *pahani*[7] register as owner cultivators. The village accountant, himself an owner of land who had been leasing out land to tenants, generally protected the interests of landowners.

With the passing of land reform legislation, by the beginning of the '60s tenancy cultivation of land owned by resident landowners came almost to an end. Only a small number of cases were an exception. When encountered with problems in evicting the tenants, the landowners took care to prevent any move by them to gain occupancy rights under the prevailing law. This was done by creating or altering the records to show that the land had been mortgaged and not leased out on tenancy. Under the Karnataka Land Reforms Act, 1974, mortgaged land does not come under the tenancy reforms.

In contrast, the situation with respect to land owned by absentee landowners was mixed in nature. Almost all such landowners were able to sell their land without any legal problems. This was possible not merely because of their foresight but also due to a lack of tenants' knowledge regarding their rights under the law. In a few cases, the tenants were offered the land for purchase. The absentee landowners with smaller holdings, say of 10 to 15 acres, or with minimum number of tenants were able to sell their land without any resistance from their tenants. Initially, during this period of mass eviction, tenants tried to gain tenancy in other lands. Further, the smaller landowners invariably approached the village leaders in all transactions, who, in turn, were capable of pressurising the tenants in the interests of the landowners. Quite often the village leaders helped such tenants who were interested in buying the land for themselves, but usually a portion of such land was also sold to others who were recommended by the leaders. In any case, by the early 'sixties almost all small holdings of the absentee landowners were sold either to the tenants or to others, or both. Tenants seeking protection under law either to prevent eviction or in getting rent fixed was altogether avoided.

Absentee landowners of larger holdings faced some resistance from their tenants when their interests were being ignored by the former. In contrast to the small holders, the bigger landowners had to deal with more number of tenants. Here again, the role of local leaders was crucial in determining the nature of transaction. A few cases will illustrate the point.

1. *Kudur Landowner's Case*

This land transaction, which took place in the mid-fifties, gave a new turn to the factional politics in the village. The Kudur landowner, a Brahman and a government servant, held about 20 acres of land in the south and south-western tract of the village site. Five tenants had been cultivating his land. All but one among them were evicted by the year 1951-52. The lone tenant had gained the confidence of the landowner, and was promised that the land shall be offered to him if and when the landowner decided to sell it.

Around this time, one of the village leaders had differences with the other leaders of the village, headed by the then *patel* and a *de facto* 'headman'.[8] The dissident leader had shown an interest in buying the Brahman's land to help his immigrant brother-in-law (sister's husband) to settle down in Rajapura. These two had already posed a challenge to the headman by trying to bring a new washerman and smith to function in the village violating the traditional *jajmani* relations with the hereditary washerman and smith in the village. These two relatives discretely approached the Brahman landowner in Kudur and came to an oral agreement to buy the land.

The tenant was a member of the headman's faction. On learning of the pact between the two relatives and the landowner, he approached the headman for help. He was suggested by the headman to offer to buy the land, and that headman would arrange the money required. The condition was that the tenant shall keep half the land for himself, while the remaining to be sold in turn to the headman. Accordingly, the tenant met the landowner and offered to buy land at the prevailing market price. On learning his genuine interest in buying the land, and having earlier committed to sell it to him, the landowner finalised the sale deed with the tenant. Exactly one month after the land transfer, the headman got ten acres of land transferred to his name, which, in due course, he sold part of it to his other supporters.

The leader had made use of this transaction to strengthen his political base and to outsmart his rivals in the village. Had the tenant been guided to seek the protection under the law, it may not have been possible to have a share in the same land for the headman, and to distribute favours to his supporters.

2. *The Case of Ramapalya Lands*

When during the mid-60s, measures of *Inam* abolition was in the air, the *inamdar* of Ramapalya sought to sell his lands to a wealthy farmer from Mandya region.[9] On learning of this, some of his forty odd tenants requested the landowner to sell a portion of the land to them. The landowner, on his part, was keen on selling the entire land in one piece to the Mandya farmer, and was aware that the farmers/tenants of Rajapura were unable to match the offer being made by the outsider. Only a few tenants were bent upon buying the land while a majority had passively accepted the eventuality. The dissatisfied tenants now approached the village leaders to intervene on their behalf with the landowner. When the landowner refused to oblige the leaders, the tenants were advised to consult a lawyer in Bangalore city. The tenants learnt that they were entitled for protection from eviction and that they could register themselves as bonafide tenants of the land for five successive years. The tenants now approached the landowner with a fresh request to sell the land, failing which they would seek legal protection.

Having sensed the hostility of the villagers by this time, the landowner consented to retrace his steps to sell the land to outsiders. Records were altered and prepared in such a manner as to show that nearly one-fifth of the land was under personal cultivation of the landowner, thereby to enable him to retain some land for himself and his family members. The remaining lands were sold to such tenants who were keen on buying the land for themselves. These tenants were mainly from Rajapura and the neighbouring village Kuppe. The landowner refused to sell his land to others who were not his tenants, particularly to the people suggested by the leaders for they were responsible in making the tenants to seek legal help. However, the leaders and others who were interested succeeded in buying the land later through such tenants who did not or could not afford to buy for themselves. Table 3.2 shows the pattern of land transfer from the Ramapalya *inam* landowner to the people of Rajapura and Kuppe. It may be seen that almost all the buyers from Kuppe, sooner or later, sold the land to the farmers of Rajapura.

In order to obtain the approval of the authorities, the landowner gave the transaction a colour that the buyers were all his former tenants and they had voluntarily come forward to buy the land. All such transfers were later ratified by the competent authorities under the *inam* settlement regulations. Such lands as were retained by the landowner were eventually sold by him and his sons to the people of Rajapura.

It was noted that most of the buyers of Ramapalya lands were immigrants who had come to settle in the village through their relatives already living there. They were able to buy the land through the good offices of the former tenants and some others who could influence the landowner. With regard to the role of the village leaders, it is evident that they intervened with a few motives. The leaders were opposed to any outsiders buying the land *en bloc*, possibly because that would pose a challenge and threat to their economic and therefore, political dominance in the village and region. The disregard shown by the landowner to the suggestions made by the village leaders prompted the latter to seek legal aid in solving the tenants' problems.

3. *The Lingayat Landowner's Case*

Both the tenants and the village leaders had taken a sympathetic stand with regard to the case of the Lingayat landowner. This person had inherited the land, an extent of 18 acres and in one parcel. The land was under tenancy cultivation. The landowner was operating a few passenger buses and was at one time a prosperous businessman. Around 1970, he met with heavy losses in his business, and therefore began to show interest in his landed property in Rajapura. His land owned elsewhere in the Taluk had all been divested of his ownership following land reforms implementation. The land in Rajapura is located in the coveted north and north-western tract with rich irrigation potentials, and was then the only single largest plot of land. It had the added advantage of being a *ooru pakkada jameenu,* the social and economic significance of which has been discussed earlier.

TABLE 3.2: Ramapalya Inam Land Transactions

No.	First Buyers	Area	Subsequent Buyers
1	Land retained by the *inam* holders	17.71	All sold to the farmers of Rajapura on the choice of landowners
2.	Kuppe's Muslim tenants	34.59	All sold to the farmers of Rajapura immediately after the first sale
3.	Kuppe Muslim tenants who retained the land	13.65	Sold or mortaged to the farmers of Rajapura over a period of time
4.	Rajapura's tenants and others chosen by the landowner	89.74	Only about ten of the twenty odd buyers were tenants, some of whom sold the land to others in the village
	Total*	155.69	

*Excludes the waste land and area under a tank

Source: Index of Land for Ramapalya and information gathered from the people concerned.

The landowner had maintained a fairly good relation with his tenants and the village leaders until 1974, when certain major amendments were being incorporated into the Karnataka Land Reforms Act, 1961. By this time he had about 12 tenants cultivating his lands.

Although the landowner had not, at this time, contemplated upon selling his land[10] it was more or less a certainty that sooner or later he would sell the land like many other absentee landowners of the 50s and 60s. Having already suffered losses in his business, he had wanted to save his landed interests. Anticipating difficulty in evicting the tenants he approached the headman for help. The headman, in turn, advised the tenants to give up the occupancy of the land. It was pointed out to them that the landowner and his dependents will be left without a source of income if the land was taken away from him. It was also pointed out to the tenants that they would have little chances of establishing their claims legally as the *pahani* records had been maintained to suit the landowner's interests. The displaced tenants were promised sharecropping opportunity on the land for growing mulberry and sugarcane with the landowner. Even this arrangement was made only after the sharecroppers signed papers stating that they were all daily wage labourers working for the landowner in his mulberry/sugarcane fields.

While all the tenants agreed not to press their claims, there were two exceptions who filed applications to the land tribunal for confirmation of occupancy rights. Both these tenants owned some land adjacent to the tenanted land. However, recently when the headman's sister's husband bought a portion of this land, these two former tenants agreed not to press their claims and promised to state suitably before the land tribunal. The tenants, however, did not have any records to prove their claim, and it is doubtful if any one in the village was willing to support their claim.

4. *The Case of Muslim Landowner*

This case involved 25 acres of land with about ten tenants, and it found its way to the tribunal for settlement. The landowner, a Muslim, was an arrack contractor living in a village about 25 kilometres from Magadi town. The land is situated on the eastern side of the village settlement. A portion of this land had access to a fresh water stream near the village Doddi. For a long time the landowner had not been showing much interest in his land nor had he cared to collect the rent regularly from his tenants. During the year 1974-75, one of the tenants, an opponent of the present headman, wanted to buy a plot of land from the Muslim landowner. This plot included a portion of which he was a tenant and another portion being cultivated by another tenant. The landowner was willing to sell the land and received a sum of Rs. 5,000 as an advance payment. The other tenant, who we shall call Kulla, protested the possible eviction. The landowner now approached the headman for help, for he feared similar protests from other tenants. The headman, initially offered to buy all the land for himself provided the landowner himself came to a settlement with all his tenants. He had wanted to prevent his political rival from buying the land in particular.

Meanwhile, the politically opponent tenant had advanced another sum of Rs 5,000, and got the land mortgaged in his favour. This turned out to be an ill advised move on the part of the tenant. When the tenant learnt that the headman was planning to buy the land, he consulted his supporters from other villages and threatened to file an application for confirmation of occupancy rights. However, on the request of the landowner, he refrained from doing so.

On learning of the pact between the landowner and the rival leader, the headman advised Kulla to file an application with the tribunal. This was also precipitated when the rival leader tried to cultivate the land which he had now 'mortgaged' but until then being cultivated by Kulla. All the other tenants followed the suit and filed applications for the confirmation of occupancy rights on the respective tenanted land. The tribunal gave its verdict confirming the claims of the tenants with the exception of the rival leader. His application was rejected on the grounds that he had been a mortgage holder, and therefore, not a tenant. Under the Karnataka Land Reforms Act, cultivation of mortgaged land does not amount to being tenancy.

In the above settlement, barring one, none of the tenants were landless. Even the tenant was only a landless by definition, because he was a dependent relative of a landowner in the village and continues to live with his relative even now. Since the former landowner/his heirs have not repaid the money to the rival tenant, he continues to hold mortgage on the land.

The above mentioned case studies and the discussion preceding to it suggest certain conclusions regarding the nature of implementation of land reforms in Rajapura. First, the local land-owners were able to evict their tenants and bypass the land reforms legislation.

They were successful chiefly because of the lack of awareness of the law on the part of the tenants, and also the fact that the records had been maintained to suit the interests of the landowners. Second, the land reform measures, as applicable to Rajapura, were only tenancy reforms, for there was none who owned land beyond the ceiling limits.[11] Third, a majority of the absentee landowners were also able to prevent their tenants from taking the matters to the authorities, which was done either through selling the land to tenants themselves or through the village leaders, or both. Fourth, the local leaders took advantage of their influence to favour their supporters, or to procure land for their relatives. In most cases they made use of the knowledge they had of the law for their own benefit and not in educating the tenants. The members of the Land Tribunal[12] were all affiliated to the same political party and faction as that of the village leaders. The sense of political solidarity between these members and the village leaders helped the latter to gain a far more powerful position in these matters. The landowners as well as tenants, who were aware of this, preferred to have the leaders on their side. As a result, many instances of tenancy disputes, both in Rajapura and many neighbouring villages were settled outside the Land Tribunals. Fifth, the cases that eventually went for settlement under the provisions of Karnataka Land Reforms Act, were the ones in which the local leaders either failed to bring about a settlement with the landowners and tenants, or those in which the landowners themselves suggested the tenants to take legal measures. Very often, the latter was resorted to so as to save the cost of registration and to ratify certain land transfers which violated the land transfer regulations. Under the law it is illegal to fragment land below an extent of three acres. But many farmers had bought land that are less than three acres, and in order to ratify such a sale deed, the buyer was asked to apply to the tribunal claiming occupancy rights. When the tribunal examines the case, the landowner selling such lands would admit before the officials that he had let out the land on tenancy and the claimant was indeed a tenant. Thus, a land transfer gets the legal sanction although prohibited otherwise. Further, the settlement under the tribunal does not involve any stamp duty, whereas a sale deed does.

Why were the Tenants passive?

With the backdrop of the case studies discussed above, I shall now proceed to examine the factors responsible for the passive acceptance of the outcome by tenants and sharecroppers. As elsewhere in Independent India, the evolution of agrarian reforms legislation in Karnataka was also gradual, beginning with *inam* abolition and appointment of various committees to study and suggest measures of tenancy and land reforms in the state. The policy formulation and their enforcement took a long time, thereby giving adequate opportunity to the landowners to evade any possible legal action. Much of the other aspects of land reforms, like ceiling on holdings were, by and large, not applicable (with the exception of Ramapalya lands). Ramapalya land came under *inam* tenure, and even that had been disposed of before the law could be implemented. Transactions that had taken place much earlier were now given a colour of the tenants buying the land voluntarily. With reference to non-*inam* land the law had enough loopholes in it that protected the interests of landowners rather than those of the tenants. The 1961 Act had stipulated that the tenant should have been a cultivator of given land for six consecutive years before October 10, 1957. In almost all cases the tenants had been changing their landowners or the land of the same landowner from year to year. Further, the *pahani* records were invariably entered showing the landowner, and not the tenant, as the actual cultivator.[13] Very often, the tenants themselves, many of whom immigrants, were on the look out for buying land, and when they were able to buy some, they had voluntarily surrendered their tenancy. Displacement of tenants was also a common feature as there had been a constant demand and competition among prospective tenants. The net result of all these was that there was hardly any tenant who had successively cultivated a given parcel of land for more than a year or two, so as to be eligible to claim legal protection under the Land Reforms Act.

A more important fact is that the tenants' awareness of their rights under the law was almost lacking. Under the 1974 Act, neither the tenants nor the landowners were permitted to employ a legal expert (lawyer) to plead their cases in the Tribunal. The presence of advocates, it was argued, would delay the proceedings. This was especially so during the early phase of land reforms, i.e., 1950s and early 1960s. A few tenants in the case of Ramapalya land had been advised to consult a lawyer, yet they had opted for a settlement outside the sphere of law. One former revenue official in the Magadi Taluk Office informed me that there had not been many tenants in the Taluk coming forward to seek protection against eviction, and that none had been from Rajapura. A former tenant of Ramapalya summed up the sentiments governing the relationship between the landowners and tenants, which indicate the passive acceptance of the situation: "It was the landowner who had given us the land when we had wanted to cultivate and grow our food. Now, if the government wants to take away his lands, let it. But we do not want the land to be taken like that. God will never forgive us for taking away land from another; after all we are all cultivators and it is not in our ethos."

It is unlikely that such a sentiment is common to all tenants. Yet they remained passive owing to their lack of knowledge regarding the law. Many such displaced tenants now think with their present awareness of the law: "If it had been now, we would not have yielded to the pressures of the landowners or others." However, having observed the local power politics and the dependence of people on such leaders, I doubt whether this statement was made with a firm conviction of being successful.

Villagers who were aware of the legal provisions made use of their knowledge to suit their own needs. Thus, when a political opponent of the headman wanted to buy a plot of land which the headman himself had not succeeded in buying, the former showed the legal path to his supporters. On the other hand, when a few tenants in another plot of land want to take to legal action to prevent their eviction, the village leaders spoke of humanitarian considerations and pointed out the legal difficulties involved in successfully establishing their claims. Indeed, if the leaders had wished it, it was not difficult for them to bring the members of land tribunal for a "spot inspection' and provide evidence to the effect that the tenants had actually been occupying the land.[14]

In recording the *pahani* details the former village accountants were, by and large, sympathetic to the landowners than to the tenants. Wherever necessary, they even fabricated the records to suit the legal requirements. The hereditary accountant of Rajapura had himself leased out his land on tenancy to a few farmers. When the hereditary offices were abolished in Karnataka, new village accountants were appointed, who were mostly from distant regions. They had no sense of involvement with the villages assigned to them. Working for a meagre salary of Rs. 100 initially and which was later enhanced to Rs. 250 per month, and assigned to villages which were socially and economically alien to them, the new village officials were eager to get back to their "regions" or home-towns. The village leaders with their links with the regional political network was successful in "punishing" or "rewarding" officials ranging from the village accountants to the Special Tahsildar for land reforms. The leaders were also kept informed of the applications filed before the land tribunal, thanks mainly to their rapport with the bureaucracy. Pressures were brought, through foul or fair means, on such applicants to 'withdraw' the applications. Having once filed an application, technically it became an irreversible record, yet it was not an impossible task for those interested to get one back. On failing to get back the application, the applicant was made to swear in the temple that he would disown his claim for the land when the case came up before the tribunal. I may mention here that many such settlement sessions (*nyaya*) held in temples were attended by officials of the police and revenue departments, and in many cases the initiatives were taken by the officials themselves.

The tenants were shy of legal process not merely because of their ignorance of legal provisions or because of the sentiments governing their relations with the landowners, but also because it would take their time away from their work, involve expenditure and they lacked support from the fellow villagers. Beginning with filing an application, till the final verdict was given, the tenants had to go to the town every day, wait till their names were called out and present their evidence. Often, their case would not be taken up on a particular day for lack of time, or because the members of the tribunal had to go for a spot inspection, or also because the tribunal did not meet for want of quorum. On all such visits to the town, the tenant has to take his witnesses, and meet the expense of all such men accompanying him. I had come across several cases where the tenant had to pay a sum of Rs. 10 every time to the accompanying witnesses as their fee. Therefore, from the cost-benefit point of view a settlement outside the land tribunal was preferable. On the contrary, land tribunal was made use of for the purpose of saving the cost of registration or to regularise a land transfer that violated an existing law.

Leaders' Perception of their Role

Why did the local leaders play the kind of role they did in the matter of land reforms? In answering this, let me first present the explanations given by the leaders themselves, although these appear to be inconsistent with their actions. The former *Patel*, who for sometime served as a village accountant following the abolition of hereditary village offices, in turn, questioned the very purpose of land reforms as was implemented by the government. I refer to this here mainly to point out the attitude towards land reforms, shared by many others in the village. He asked me: "What is wrong in a landowner leasing out his land to tenants? Like any other person he too wants to get returns from his ownership of the property. If the government wants to distribute land to landless or poor people, let it give on its own, but not by taking it way from some others." He then pointed out that "even when the land to the tiller programme was carried out, the beneficiaries were mainly the people who owned land". From *Patel*'s point of view, land reforms had been merely transferring one's land to another. This comment by *Patel* holds good, at least in Rajapura where almost all the tenants who were successful in obtaining the occupancy rights, or those who bought land from their tenants were landowners.

The inconsistency in *Patel*'s attitude and actions becomes evident if one is informed that he too is an unsuccessful applicant for confirmation of occupancy rights over a plot of land which his undivided family enjoyed as priests of a local temple. When asked about this, he gave an evasive reply saying that he had applied because he did not want to take chances with the unpredictability of future legislation for such temple lands.

The "headman" of Rajapura, on the other hand, did not have much to complain regarding the aims of and need for land reforms. He rationalised his role as having been guided by the merits of individual cases. Referring to the case of the Muslim landowner[15], he stated that he had wanted "as far as possible, the tenants to be given an opportunity to buy the land without having to go to the tribunal. But, if the landowner wanted to displace the tenants or wanted to sell the land to others — especially when the tenants wanted to buy the land for themselves — we had to suggest the legal course of action to the tenants." He also pointed out some instances among the former tenants of the Muslim landowner who even now sent or gave some quantity of grain (in lieu of rent) even after the settlement was made through the land tribunal.[16] With reference to the Lingayat landowner[17], he justified his role by stating that it was an exceptional case where the landowner was resuming land for personal cultivation. "If the landowner was trying to sell the land to non-tenants or to outsiders by evicting the tenants, our stand would have been different in the case" he said.

These two leaders seem to subscribe to the views that the landowners should give an option to the tenant to buy the land before looking for another buyer, and that the landowner is welcome to resume land for personal cultivation. However, by looking at the transactions in most other cases, it becomes evident that the "merits of the case" which guided the local leaders were mainly determined by their personal equations with the landowner, tenants, and other people involved.

One finds a pattern in the personal relations of the leaders with the individual tenants who were able to buy the land, or with those who went to the tribunal, or with those who were able to buy the land, or those others who were able to buy land without having been tenants. The leaders usually took a favourable attitude towards the tenants or the buyers, or even the landowners, if they met the approval of the leaders, which, in turn, depended upon many other considerations, namely, who were the people in terms of caste and kinship, whether or not they were of the same faction, and so on. This can be seen in the case of the Muslim landowner discussed above. There is a keen competition in the village to buy or control land between the two factions headed on the one side by the headman and the former *Patel*, and on the other, by the opponent involved in the Muslim landowner's case. Both the *Patel* and headman had initially wanted to buy the Muslim's land, mainly to prevent their opponent from buying it. One condition was that the landowner himself should evict the tenants through an amicable settlement. But, when the landowner could not succeed in doing this, and when he sought to sell the land to others, the leaders suggested legal course of action to the tenants. In this process, one sympathiser of the other faction also gained. The point to be noted here is that the leaders did not suggest the legal course of action to the tenants when they themselves were contemplating upon buying the land.

Similarly, in the case of the Lingayat landowner it was no accident that the leaders took a sympathetic attitude. It was the only single largest plot of land in recent years. It had the added advantage of being on the north-western tract and being adjacent to the village settlement. Almost all the bigger landowners had a desire to buy this land. A few, including the *Patel*, his brothers, the headman, and his brother-in-law, as well as the political opponent of the headman were the ones in the village who could afford to buy the land. But they all wanted the land to come to the market on the landowner's initiative so that they could buy the land on their own terms. They were also aware that if the tenants succeeded in claiming their share of the land, the fragmented land loses its value, and then it would be difficult to buy from many rather than a single landowner. The aspirant landowners were certain that it was a question of time before the Lingayat sold the land. Adding to such speculation, the landowner himself had admitted that he may have to sell the land to meet the marriage expenses of his three daughters.

It was, therefore, no accident that none of the tenants had such relationship with the leaders that warranted a suggestion and help in taking legal aid to press their claims for land and to gain protection against eviction. When one tenant made an application for confirmation of occupancy rights, more so because of the pressure from the other faction, a counter application was filed by another tenant for the same plot of land to make the legal process complicated. However, neither had any documents to support his claim.

As it turned out during the years 1979-81, the land came into market. The young headman was offered the land for buying. Having married a few months earlier and contested an election the year before, the headman was not in a position to raise the money to buy at such a short notice. A portion of the land has now been bought by the headman's brother-in-law, who is an engineer with the Karnataka Government. For some time, cultivation of this land was being supervised by the headman's brother. The remaining portion of the land was bought by *Patel*'s brother. The two former tenants who had applied for confirmation of occupancy rights had agreed not to press for their claims.

It thus becomes evident that the role played by the village leaders was determined, above all, by their personal and kin interests, and that of their faction. It was not tenancy reforms to tenants *per se* which determined the merits of the cases, but who the tenants or buyers were, and whether or not the leaders themselves were interested in buying the land. When this conclusion was pointed out to the two leaders in a lighter vein, they commented that "that is how it has turned out to be in the final analysis."

One may ask: why this special emphasis upon the role of leaders? The answer to this also brings us to the consideration of another question: how were the leaders able to play their role as noted above? Earlier I referred to the relation of the leaders with the officials and politicians at the taluk and district level, which explains in part their ability to manipulate events at the village level. Secondly, the villagers, whether tenants or buyers of land, were invariably dependent upon the leaders for counselling in all matters concerning not only land but also money, loan, market, and matrimony. In the case of tenancy claims, during the course of obtaining and filling up an application, submission of it on time with all the relevant documents, obtaining all such documents from various offices in Doddi and Magadi, the tenants always needed the help of the leaders of either faction. To pursue a course of action without the guidance and support of the leaders meant readiness to face all consequences on their own. The leaders had a greater experience in these matters, and knew where to get certain kinds of information, whom to approach for certain kinds of favours, how to get things done faster in the revenue or other offices, and also how much money to be paid as bribe to the officials.[18] Further, it was not uncommon for the officials on their own to enquire with such people who approached them directly: "where is your leader/ headman? or "what do your leaders think of this?" In some cases the leaders requested the officials to adopt delay tactics or to misguide the people so as to manipulate the legal process for their own benefit or to help their supporters. Added to these difficulties there was also an important notion prevalent among most of the tenants. One of the four non-official members of the land tribunal was a regionally dominant politician, of whom the local leaders were strong supporters. The other members of the tribunal were also members of the same political party supported by the headman's faction. The tenants feared that if they went to the land tribunal on their own and without the approval of the leaders they may not succeed, because of the political links of the local leaders with the members of the tribunal.

Officially, it is claimed that is not possible to effect individual biases by the members of the tribunal. One official claimed that since the tribunal consisted of four members and one ex-officio member, it was not possible for the members to be biased.[19] However, it is argued here that it is possible for one member to influence the other members, specially because they all belong to the same political party or are proteges of the same politician appointing/nominating them to the tribunal. A few members of the tribunal and a few officials of the land reforms cell at the taluk level admitted the possibility of yielding to prejudices and pressures. It was a common sight to find the houses of tribunal members being crowded by people or leaders, tenants, applicants, landowners, etc., from other villages, seeking help.[20] In complicated cases, the tribunal members go for an on the spot inspection in which the biases of the members and of the local leaders play an important role in twisting the case as per personal prejudices.

As a result of all these reasons, a few tenants who had contemplated upon taking legal course of action, even in the absence of support had given up the idea. The leaders, therefore, become vital in all matters of tenancy reforms who acted, as they claimed, upon the "merits of the case".

Beneficiaries of Land Reforms

.One of the slogans employed with regard to land reforms has been "land to the tillers", according to which the beneficiaries ought to have been tenants and the labourers. Many studies, however, point out to the contrary (see, e.g., Joshi 1979; Herring 1980). Let me now proceed to examine who the beneficiaries of land reforms in Rajapura were. Speaking in terms of definition, the beneficiaries ought to be the tenants, who having been tenant-cultivators, obtain the occupancy rights and subsequently by paying the premium[21] and land revenue, will become owners of the land. It implies, then, that the 'beneficiaries' are the ones who obtained the occupancy rights or ownership rights through the Land Reform Act. In Rajapura, I have to distinguish between those who first arrived at a settlement and later formalised it under the Act, and those who had the settlement only through the legal process. Almost all the buyers of Ramapalya land and that belonging to the absentee landowners may be grouped under the former category, whether or not they were genuine tenants of the land.

The initial settlement of Ramapalya tenanted land was done with regard to 16 buyers from Rajapura, involving 89.74 acres of land; nine buyers were Muslims from Kuppe village involving 48.24 acres; and remaining 17.71 acres of land was retained by the landowner and his family members. Among the buyers from Rajapura, 56 per cent were recent immigrants, who among themselves bought about 46 acres. Further, among the immigrants who bought the land, 56 per cent were already owners of land in Rajapura, while the remaining were landless. Some of these landless immigrants had come to settle in the village under the practice of *manevalthana*, in which the son-in-law begins to manage his father-in-law's household. Ultimately, the daughter inherits the property. Thus, many such 'landless ' had landed interests in this manner. Many among the landless immigrants were in fact former landowners prior to migration, who had sold their lands and were waiting to invest in land in Rajapura.

Similarly, in the more recent case of Muslim landowner's tenants, the beneficiaries were, by and large, landowners. A few had owned more than ten acres of land, while one owned as many as 22 acres. It can, thus, be seen that most of the beneficiaries were landowners themselves (see Herring 1980, for a similar finding in Kerala).

Land Transactions in the Wake of Land Reforms

Given the nature of implementation of land reforms, what is its impact upon agrarian relations in Rajapura? As one can foresee, there may result a contraction or disappearance of tenant-landowner relationships, in which the tenant is either evicted or is able to obtain the occupancy rights over the land, and perhaps ownership rights. Consequently, the evicted landless tenant may become a landless labourer, or if he had held any land of his own, will remain a mere land-owner after eviction. A successful tenant after the reforms becomes a landowner or a bigger landowner than before (see Figure 3.1).

In order to examine the impact of land reforms in the village, I have identified such land transfers which took place between 1950 and 1978, involving tenancy cultivation, eviction of tenants, or tenants buying the land or gaining occupancy rights through the reform legislation. This data have been gathered from the residents of the village, and in a few cases, from absentee landowners who sold the land, now living in Magadi town, Bangalore city and other places. The data presented in Table 3.3, has been quantified by adjusting the incidence of partition of the family property. Further, the landowning or landless status of the individuals as recorded here pertains to his status at the time of transaction. In some cases, the tenant may have been evicted by one landlord, but if the same tenant had been able to buy land from another landlord of whom he was a tenant, the latter has been taken into consideration. Finally, the cases referred to as "evicted" pertain mainly to the eviction by absentee landowners, although there are a few evicted tenants of the local landowners, whom we could identify as having been tenants for more than 3-4 years.

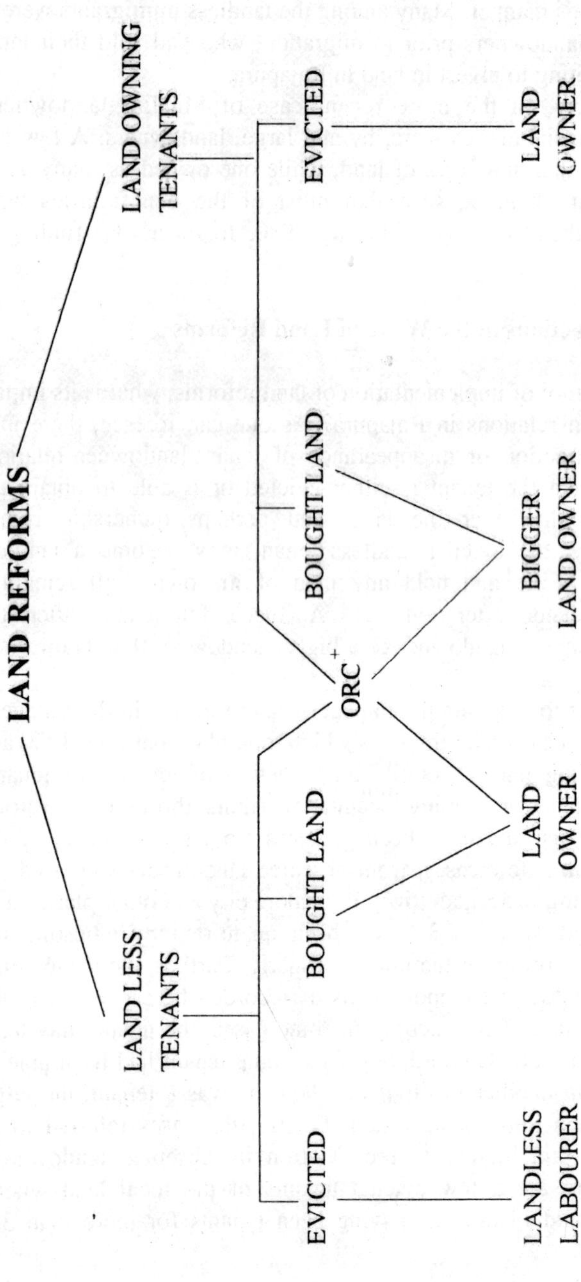

+ *O.R.C. Occupy Right Confirmed, under section 48 (a), Karnataka Land Reforms Act, 1974.*

FIG. 3.1: Outcome of Land Reforms

TABLE 3.3: Land Reforms and Land Transactions (1950-1979)

(in percentages)

Cultivator's Status	Evicted	Bought for self	Bought to sell	Bought partly self/ selling	O.R.C.* Pending	Total	
						No.	%
Landless tenants	66.7(10)	13.3 (2)	–	6.6 (1)	13.3 (2)	15	100
Landowning tenants	37.2 (16)	9.3 (4)	4.6 (2)	11.6 (5)	37.2 (16)	43	100
Total	44.8	10.3	3.4	10.3	31.0	58	100
N	(26)	(6)	(2)	(6)	(18)		

* Occupational rights confirmed.

Note: Figures in parentheses refer to number of cases.

Tables 3.3 and 3.4 together indicate the impact of land reforms upon the two agrarian categories: the landless and landowners. It is seen that a majority of tenants (74.1 per cent) were also landowners (see Table 3.3). This is in keeping with the fact that landowners generally seek tenants from among the landowning cultivators who not only have a skill but also capital, cattle, manure, and agricultural implements to cultivate the land efficiently, but also are able to withstand the vagaries of the climate so as to make regular payment of rent. In the "evicted" category (see Table 3.3), a majority were landless tenants, most of whom were immigrants. Their social status was also

TABLE 3.4: Buyers of Land in the Wake of Land Reforms (1950-1979)

(In Percentage)

Non-tenant buyers	Bought for self	Bought for self and partly to sell	Total	
			No	%
A. Immigrants				
(i) Landless	94.7	5.3	19	100.0
(ii) Landowners	100.00	—	1	100.0
B. Residents				
(i) Landless	—	—	—	—
(ii) Landowners	33.3	66.7	9	100.0
Total	75.9 (n) (22)	24.1 (7)	29	(100.0)

lower since most of them were of non-Okkaliga castes like Lambani, Odda, Holeya and Madiga. Some of them had left the village during 1957-58 and again in 1962-64, owing to a drought. Nevertheless, eviction of landowning tenants did occur in a few cases where the landowner was keen on buying the land but could not afford to compete with others.

As pointed out earlier, it was the others rather than the tenants who were able to buy land in the wake of land reforms. Many such people as well as those who had made use of the land tribunals to regularise the transactions made earlier, constitute the category under "O.R.C". Further, if we exclude those who bought lands in order to sell them either in part or fully to others, (10.3 and 3.4 per cent respectively), the above statement appears to be even more convincing. Many such tenants bought land, even if to sell partly to others, by depriving other tenants of their right to cultivate it or an opportunity for them to buy it. Finally, the proportion of tenants buying the land is higher among the landowning tenants than the landless corresponding with the incidence of eviction being higher among the latter. Similarly, the proportion of tenants who resorted to legal course of action is higher among the landowning tenants than the landless tenants. These last two observations have to be viewed in the light of the fact that it is dependent on the very access tenants have for land to cultivate, capital to buy land, and above all the support of the local leaders.

Finally, I shall examine the instance of non-tenants buying land which had been under tenancy cultivation, therefore, causing eviction of tenants. Table 3.4 presents 29 such cases, in which the buyers of the land were not themselves tenants. The land belonged mainly to the absentee landowners, and all these had been sold in order to evade land reforms legislation. A distinction between 'immigrants' and 'residents' has been made to show the extent to which the availability of land attracted people from outside the village to settle in Rajapura, and also the extent to which local tenants had to face displacement on account of such immigrants buying land. There are instances in which a buyer had caused eviction of more than four tenants. In some other cases, the buyers had to fabricate records to ratify the purchase through land reforms legislation.[22]

It can be seen that more than two-thirds (69 per cent) of the buyers were immigrants. With the exception of one, all of them were landless at the time of buying the land. Among them only one had bought land to partly sell to others. Further, it may be noted that no landless resident bought the land. In contrast to the 'immigrant landless' buyers, only a third of the resident landowning buyers bought the land exclusively for themselves, while the remaining bought it to sell in turn to others, mainly to their relatives and sponsors.

Impact of land reforms becomes clear if one considers the observations from Table 3.3 and 3.4, keeping in mind the pattern presented in Figure 3.1. While a few landless tenants were evicted, a still larger number of landless immigrants were transformed into 'landowners" as a result of the transactions that took place in the wake of land reforms. As I shall point out later,[23] the 'landless tenant' turned 'landless agricultural labourer' began to have opportunities of becoming a sharecropper or tenant in the coming years. I may, however, note for the present that the change in the status of 'landless tenant' into 'landless labourer' was indicative of a reverse mobility process for him. As noted earlier, such landless tenants or members of their families were also working as agricultural labourers even though they were tenants. There were several instances where one or two sons had been attached as bonded labourers, and the female members working as wage labourers. Hence, neither their becoming tenants, nor later downward mobility as labourers had any significant impact upon their social and economic status. Eviction of landowning tenants, on the other hand, did not alter their social status significantly, in which the change was from being landowning tenants to now being merely landowners. As noted earlier, the need for tenancy was mainly to procure land for their immigrant relatives or other sponsored tenants, rather than for the income to be derived from tenancy cultivation for themselves. In some cases, the displaced tenants had now surplus labour, or had to seek additional source of income in the absence of tenancy cultivation.

When a landless tenant bought the land, or more recently, obtained occupancy rights, a change occurred: he now became a landowner. One of the results of this transformation was that he need not give a portion of his produce as rent to the landowner, and therefore, had a higher income for himself. Among the three landless tenants who acquired land by buying or through reforms, only one of them (Case 1, above) had a marked change in his life-style. He had now an adequate income from the land to support the education of one of his sons. The family had no history of having worked as labourers either before or after buying the land. He did cultivate land as tenant of another landowner for sometime, a portion of which also he bought later. Initially, the household had to raise a loan in order to buy the land, which was advanced by the headman in return for the favour of selling part of the land to him. Another landless tenant, an Odda by caste, had at one time been a bonded labourer. Later, he secured tenancy with part of Ramapalya land, but was soon evicted. Later, having been a tenant of another plot of land, he was successful in buying it. Nevertheless, even after buying the land the three landless tenants had continued to be tenants of other landowners until they bought some more lands. Although in an indirect way the coming of land reforms helped these three landless tenants, there had been a greater number of displaced tenants who were reduced to being agricultural labourers.

For the landowning tenants, the impact was that when they bought the land, they became bigger landowners. Nearly 60 per cent of the landowning tenants were benefited in this manner. The implications are that, first, those among such landowners who were occasionally working as agricultural labourers could now afford to engage in merely supervising the work of agricultural labourers or employ bonded labourers to meet the increased work on fields, or rely upon exchange labourers. Secondly, the transfers of land, by and large, strengthened the hands of the already well-to-do farmers than the poorer ones.

With respect to the incidence of people buying land thereby causing the eviction of other tenants, (see, Table 3.4), it may be said that the landless immigrants who bought the land not only became landowning immigrants, but also gave a market for the local labour supply. In some cases, the evicted landless tenants themselves had entered bonded service with such landowners. Further, the coming of immigrant landowners and transformation of some landless tenants as landowners also gave a new lease of life for the traditional *Jajmani* relations, concerning both the ritual and secular services. For example, there was now service relations between smiths, barber etc., for all the new landowners, while as landless tenants they were ineligible for such services.

Continuities of Traditional Agrarian Relations

It is true that as a result of the absentee landowners selling away their land or losing their land through land reform legislation, there was a contraction of tenant-landowner relations. But the tenant-landowner relations having thus minimised for a while, began to reappear gradually, and in some cases, it may be attributed to the very land reforms itself.

During the years 1975-76, there were some instances of 'tenants' applying to the land tribunal for confirmation of tenancy/occupancy rights. There was one particular case involving the land belonging to the temples in the village. There are two temples with Lingayats officiating as priests. Each of the two temples has some *kodige* (*inam*) land, which was enjoyed by the officiating priests. The offices of the priests are not hereditary, for no Lingayat household has lived for more than 20 years in the village, owing to, among other reasons, the Okkaliga dominance. In 1975, one of the priests made an application for confirmation of occupancy rights over the land of the temple. It is also said that the priest was a sympathiser of the faction opposed to the headman and *Patel*, and that he was induced by the other faction to apply for rights over the land.[24] Thanks to the information network that the headman maintained with the bureaucracy, he was informed of the application. The headman and his faction was successful in getting the application 'withdrawn'.

Following this incident, it was decided that the priests were not to be entrusted with temple lands. The priest in the meantime left the village and in 1976 a new priest was appointed. Only a portion of the temple land was given to him for his use. Thus, out of about 12 acres of land belonging to the temple under question, only about 5 acres, and from the other two temples 7.5 and 2.5 acres each were earmarked for the priests. The remaining temple lands were decided to be auctioned for a term of one year regularly. Such land was divided into about five smaller plots and the bidder offering the highest rent in cash was given the right to cultivate the plot for one year. Thus, the fear of land reforms now gave room to a kind of institutionalised land leasing. I may also mention here, that the tenants have not been regular in paying the rent, while the leaders of the opposition faction accuse their opponents of embezzling money. The newer tenants in subsequent years refused to pay the rent until they were shown the accounts of income from the leasing out of temple lands in the past. Now that the collection of rent has not been regular, auction in the previous two years witnessed severe competition. Big landowners were found to be bidding huge and impossible sums for the rights of tenancy. One former untouchable landless remarked: "Because they know that they do not have to pay anything, they are bidding sky high rents. If the same was leased out to landless like us, we should have been benefited. They will not do so, for if they did, who will come to their houses to work and 'split the ragi balls'!"[25]

There is another important way in which tenancy and sharecropping persists in the village. Many people owning property in the undivided families now live in other places like Bangalore city, Tumkur, Ramanagaram, etc., engaged in salaried non-agricultural occupations. Often their earning is also invested in buying the land in Rajapura. Such land may have been cultivated by the members of the undivided family. However, when the family property is divided among brothers, such non-resident brothers invariably demand a share in the landed property. It is rare that a brother, who comes out of a joint family, is on good terms with his other brothers. As a result, the land has to be entrusted to members outside the family for tenant cultivation or sharecropping. There were more than ten such cases in Rajapura, while a few others sharecrop with brothers. Proper care is taken to ensure that the tenant does not make a claim for rights over the land by checking personally with the village accountant, and by changing the tenants as often as possible.

A major reason for the revival and continuity of sharecropping and tenancy is the introduction of a new agro-based commercial activity, namely sericulture. Its technology requires intensive cultivation for better returns, and a greater need for labour. In the following chapter, I shall examine the nature of changes that occurred as a result of the introduction of sericulture and note how there is continuity in some other aspects of agrarian relations.

Notes

1. *Inam* tenure refers to such land, sometimes an entire village, which were exempt from the payment of land revenue either wholly or partially (*sarvamanya* and *jodi* respectively). See, Rao (1929) for a description of different forms of land tenure in the princely Mysore state.
2. e.g., the accountant, headman, village messenger, etc.
3. The Mysore Agricultural Land Laws Committee. Mr. B.D. Jatti was its Chairman, and therefore, the committee is known as Jatti Committee.
4. The new state incorporated some parts of Maharashtra, Andhra Pradesh, Coorg and the former Madras presidency, all of which had different tenurial conditions having been under different administration earlier.
5. For a further discussion of evolution of land reforms in the Post-Independence period

in Karnataka, see, George (1970), Rajan (1979), NCA Report (1976), Manjappa (1984), Pani (1983) and Ranganathan (n.d.)

6. See also, Henchette (1972) and Manjappa (1984).
7. A register maintained by the village accountant to record the cultivator's name, owner's name, crop raised, source of water, etc., for all land in the villages.
8. While the *Patel* was the official headman of the village, this 'headman' was a confidante of the *Patel*, and in subsequent enjoyed greater influence and power in the village.
9. Such efforts were in keeping with the general trend concerning *inam* lands all over the state and elsewhere. See, for e.g., Hill (1982:260) for a similar process in another Taluka in Bangalore district, Karnataka.
10. During 1979-81, and again 1982-83, the landowner and later, his heirs sold all the land.
11. The Land Reforms Act, 1974, reduced the ceiling on land holdings from 216 acres to 54 acres per family for dry lands, and from 18 to 10 acres for wet lands. In the state as a whole, 1,28,583 acres of land was determined as surplus lands and about 44,500 acres (34 per cent) of land was redistributed. Further, out of the 80,03,55 tenancy cases that came up before the land tribunals under the 1974 Act, 41.7 per cent had been disposed of in favour of the tenants involving 12,89,319 acres of land by September 1980 (Ranganathan n.d.).
12. It is noteworthy that all the members of the land tribunal in Magadi during my study were landowners. One among them, a former MLA, had nearly 35 tenants who had applied for confirmation of occupancy rights over the lands. This member's large estate of over 350 acres had been transferred in the names of his brother, sisters, brothers-in-law, etc., as early as 1961 so as to evade ceiling in future. All these "holdings" continue to constitute the estate as it had been in the past, personally managed by him.
13. In the state of Bihar, in 1949, the *zamindars* had controlled not only 90 per cent of land, but also had all the records under their control. The state had very little knowledge of who owned the land and who were the tenants (see, Sinha 1978).
14. A former Chairman of the Land Tribunal in Magadi informed me that the favourable evidence obtained during a "spot inspection" was enough to hold a decision in favour of the tenant even if the land records were to the contrary.
15. See case No. 4 above.
16. This was confirmed by a few former tenants of the landowner, who said that they gave the foodgrains to the landowner not only because the former landowners were now poor, but also because "every landowner has strong attachment to his land and likes to have some foodgrain grown on that land." Many other former tenants who had bought land from their landowners even now make gifts of foodgrains, fruits and vegetables grown on such land, once or twice a year to their former landowners who have now settled down in cities. Indeed through such relations the farmers maintain and extend their urban contacts.
17. See case No. 3 above.
18. It is said that the officials were usually satisfied with the minimum offered as bribe through the leaders of headman's faction.
19. Based on personal interview with Shri M.A.S. Rajan, the then Special Adviser, Land Reforms, Government of Karnataka. See also his work (Rajan 1979:19) in which he states that the statistical probability of a member indulging in a biased decision is only 50 per cent, and that the likelihood of a four-man tribunal taking an unanimous prejudiced decision is one-sixth or 16.7 per cent.

20. Membership in the land tribunal became important socially, economically and politically during the years 1977-80, and even later. The members are said to have amassed wealth by virtue of their being members of the tribunal (Manjappa 1984; Nataraj 1980).
21. The payment of premium was subsequently waived for small and marginal farmers.
22. See case No. 2 pertaining to Ramapalya lands.
23. See Chapter 4.
24. The leader of this other faction is a Lingayat living in the neighbouring village Doddi. Since 1987 there has been a truce between the two faction leaders.
25. *Avara mane hittu muriyoke yaru bartare?*, meaning who will be bonded labourers in their houses and eat ragi balls they give us ?

4

Under the Impact of Sericulture

Closely on the heels of land reforms, Rajapura and the region began to experience a shift in the agricultural technology with the introduction of sericulture. In this chapter, I shall discuss the changes that took place as a result and show how some of these changes reflect continuity of some of the pre-land reforms agrarian the relations. First, I shall briefly discuss the level of monetisation in agricultural economy.

Prior to 1970, income of farmers by way of selling their produce was minimal both in terms of extent and frequency. A typical farmer allocated almost all his land to the cultivation of ragi, the staple foodgrain. As cultivation of ragi permits mixed cropping, he also raised other crops like jowar (mainly as fodder), mustard, castor, black til, pulses like horse gram, cow pea, pigeon pea, and most importantly *avarekai* (field beans) — all referred to as *salu bele* (row crops). Depending upon the extent of land held by a farmer, he also grew groundnuts and some vegetables on a separate plot. In a sense, all these 'row crops' are cash crops, for they were/and could be sold in the market. However, the quantity actually marketed, including ragi, was and is still determined by (a) size of land on which the crop is grown (which in turn is determined by the size of land owned or leased in); (b) kind of produce it is, i.e., its marketability and durability, (c) its requirement for the household's consumption; and more importantly, and (d) the need for money.

Viewing from the present situation, the pattern of landholding is not significantly different from the past in which a majority were small and medium sized farmers.[1] Tenants and sharecroppers had to share their produce with the landowners. Out of the farmer's share came, if any, marketable surplus. It was only a few bigger landowners with more than 10 to 15 acres, and the firewood gathering agricultural labourers who had a more frequent cash income than the marginal and small farmers. Referring generally to all the farmers in the village, a majority of them retained more than three-fourths of ragi production, selling the rest only if they had sufficient surplus. It is argued that farmers have two sets of 'imperatives' determining the marketable surplus, which Wolf (1966) calls the need for ceremonial fund and funds for rent. The need for ceremonial fund occurs when men are "largely self sufficient in food and goods production, (and) they must entertain social relations with other fellows" (Wolf 1966:7). These are, for example, hosting dinner to reciprocate hospitality of others, for marriages, births and deaths. Farmers also had to meet the grain requirement to make payments in kind to their labourers and *servicing castes*. The rent fund is the surplus earmarked for payment in kind to the craftsman and agricultural labourers, and in case of tenant cultivation, the share of landlord, or the land tax.

Farmers sold some of their produce in order to meet the expenses arising out of a forthcoming event, like marriage, or to have cash to pay for the labourers for harvest operations. All the requirements for a marriage, as for instance, ornaments, clothes, travelling expenses, etc., could not have been otherwise met with, given the kind of produce they have either now or had in the past. Hence, the ceremonial fund, and partially the rent fund, had to be converted into cash in order to meet such expenses. A farmer also sold a larger quantity of grain than usual in order to repay a cash loan or to make a down-payment for buying land.

Both in the past and the present, some products have occupied an important position in being a source of cash income. They are *avarekai* and groundnuts. The latter is only by farmers who grow it in more than 0.20 or 0.30 acres. The green *avarekai* is sold as a vegetable, which is very popular in the urban areas of Karnataka and elsewhere in South India. Dried *avarekai* is sold as pulses. Unlike any other vegetable, therefore, it is not a perishable produce, for it has its value as both vegetable and pulse. Rajapura farmers travel to Ramanagaram, Magadi or Bangalore city to sell the produce during the season. While no bus or public transportation was available during a major part of the year, a bus plied from Magadi to Ramanagaram through the village once a day during the months of November-January; the incentive for this being to carry men and sacks of *avarekai* to Ramanagaram and Magadi. On an average, from one acre of ragi field, on which *avarekai* is grown, a farmer picks as much as ten to fifteen sacks in a month, (approximately, 20 to 25 sacks during the entire season). A sack containing about 30 kilo grams of avarekai was sold on an average at Rs 25 to 35, which, however, varied depending upon its quality and on its demand and supply in the market. An acre of dry land yielded on an average a net income of Rs. 700 to 800 in a season during 197-79. The actual cost of growing *avarekai* works out to be small since it is grown along with other crops in the ragi field. A wage of Re 1.00 was paid for half a day to the labourer for picking *avarekai*.

Many farmers with smaller quantity of *avarekai* to be sold, did so to the local middlemen or the visiting merchants from Magadi or Ramanagaram. Some such buyers are on the pay-rolls of the wholesale merchants in Ramanagaram town, who meet the cost of transportation and intial capital to the middlemen. In passing, it may be mentioned here that some wholesale buyers of Ramanagaram *mandi* generally approach the farmers of Rajapura for capital during the *avarekai* season, the loan to be repaid at the end of the season with interest (usually at 10 per cent per month). In the wholesale market (*mandi*) sacks of *avarekai* are sold by auction, by a simple and crude assessment of the quantity and quality of the contents.[2] The market at Ramanagaram functioned almost like an exchange market in which the so called wholesale merchants were super middlemen. In addition to engaging men from different villages as their agents, they also engaged young boys to bring farmers coming from different villages to their *mandi*. These boys waited near the town bus-stand and on the main routes into the town. As soon as they sight a farmer alighting a bus with his sacks of *avarekai* boys from different *mandis* flock the farmer persuading him to come to their respective *mandis*. Having contracted a farmer and obtained his produce the mandi merchant often sells it to other merchants — retail or wholesale— immediately. It was observed that farmers had some particular wholesale merchants with whom they traded regularly. Such merchants do not pay the farmers on the day *avarekai* is sold to them, but during a subsequent visit by the farmer to sell the next harvest of *avarekai*. Sometimes, money is also sent through other farmers who visit mandi on a subsequent day. Delayed payment is made usually when there is a good supply and a low demand for *avarekai* in a year. On the contrary when there is a good demand but not as much supply, the mandi merchants hire other middlemen, who are often from different villages like Rajapura, to procure the produce for sale.

Apart from *avarekai*, farmers also sell some quantities of oil seeds,[3] such as castor seeds, til, pulses and groundnuts, and sometimes fodder. In addition, sheep and cattle rearing has been an important activity among some farmers of Rajapura during the past, thanks mainly to the forest in the neighbourhood and vast waste land at the foot of small hillocks within the village boundaries. It is said that about 30-40 years ago, the cattle wealth of the village was one of envy to the others in the region. A few former bonded labourers recollect that many of the landowners themselves were sometimes unaware of the exact number of cattle in their cattleshed, for there were too many, of which the bonded labourers were in exclusive charge. There used to be a sense of competition among many farmers to maintain large number of cows and bullocks in good health and fitness.

Over a period of time, the village grazing land has either been encroached upon or brought under personal holding and cultivation through government grants. Moreover, the forest officials do not now permit indiscriminate grazing of animals in the forest. The forest official — *the guard sahib* — collected, a practice even now continued, Rs. 3.00 per sheep per annum for his permitting to graze animals,[4] and occasionally felling certain trees to make agricultural tools. The advent of sericulture and a relatively higher enrolment of young boys and girls in the local schools also have been responsible for declining interest in animal husbandry. Farmers now devote more time to sericulture along with their regular agricultural activities than to grazing animals. The school, on the other hand, draws quite a good number of children who were otherwise assigned the task of grazing animals.

Nonetheless, sheep and cattle rearing continues to be an important source of income to the farmers by way of their sale and the sale of manure if they have a surplus of it. It may be seen from Table 4.1 that about 27 per cent of the households in the village were engaged in sheep rearing during the study. The per cent of households rearing sheep/goats increased along with an increase in the size of landholding with the exception of the very big landowners. Cows in Rajapura, as in the region, have an utility value both for manure and their use in agriculture, rather than dairying. Not many farmers sold milk during 1978-79, with the exception of those who had good milch cows and buffaloes.[5] Production of milk is affected by employing cows for ploughing, etc.

Almost every farmer visits one or two cattle fairs held annually in and around Magadi taluk. The famous and important ones for the farmers of Rajapura are held at Magadi, Sugganahalli and

Shivagange. Here, they sell their old pair of cows or bullocks to buy a new pair, a routine which is referred to as "exchanging of cattle". In the process a farmer may gain or lose. He may lose by way of paying more to the new pair than what the old was sold at. Sheep and goats are sold at the weekly market in Magadi held on Fridays. During 1978-79, a well grown ram fetched anywhere between Rs 350 and 500 or even more. Farmers generally sell one or two rams at a time to meet some expenses, but when they have to invest heavily or spend on marriages and such other ceremonies they sell more than ten or all sheep and goats. They also get paid Rs.50 and more for the skin of a sheep, or Rs. 50 and more for selling wool of five sheep apart from receiving a blanket.

TABLE 4.1: Landownership and Animal Husbandry

Landownership (in acres)	Goats and Sheep %	Buffaloes/ Cows %	Bullocks %
Landless	4.65	23.65	—
Below 2.50	19.05	47.62	9.52
2.51-5.00	24.29	87.80	24.39
5.01-10.00	44.44	91.67	63.89
10.01-15.00	77.78	100.00	100.00
15.01-20.00	57.14	100.00	100.00
20.00 +	40.00	100.00	86.00
Total	26.78 (n=49)	65.57 (n=120)	28.96 (n=53)

Farmers also gave their fruit bearing trees like mango, tamarind, jack, etc., on lease to fruit merchants of Magadi, Ramanagaram or of the neighbouring village Kuppe. Generally, such merchants visit the village during the flowering season and contract a few trees for a fixed sum. The merchants, however, agree to give a basketful of fruits from each tree to the farmers. Since mango trees flower during the time when the farmers begin to make arrangements to harvest ragi, they are in need of money and almost all the farmers having fruit bearing trees sell the produce to the merchants. Since tamarind is not a perishable crop, farmers generally resort to retail selling of processed tamarind throughout the year in smaller quanti-ties. Tamarind was sold at Rs. 20 to 25 per basket during 1978-79.

For more than fifty years the village has been having a visiting merchant from Magadi who buys virtually everything from the farmers. He owns a grocery store in the town. On his arrival to the village, farmers enquire the current price for the items for sale, and if satisfied with the price quoted they sell the goods. Farmers try their best not to let him know of their need for cash, lest the merchant exploits the situation to his advantage. Almost all such selling is done in small quantities, and farmers prefer to sell it to the visiting merchant rather than selling it themselves in Magadi town or elsewhere. Further, since the difference in price in the market and the one quoted by the merchant is only marginal, it would be preferable to sell goods to the merchant than having to incur expenses on transportation, wages to labourers carrying the head-load, or cost of hiring a bullock cart, and save the time. From the merchant's point of view it is profitable as he buys goods at a comparatively lower rates. It is also common practice for him to speculate on some goods.

It is important to note that the visiting merchant too has found a place in the network of exchange of goods and services in the village under *jajmani* relations, locally known as *adade*. He received two bundles of straw annually from all the farmers with whom he had business. This is usually soon after the harvest, in the same manner as did the other service specialists. I mention this only to point out to an extension of certain aspects of traditional patron-client relations to the set of relations concerning the visiting merchant with whom the farmers have mainly a monetary relationship.

The Significance of a Cash Crop

It may be seen from the foregoing account that the village's economy prior to 1970s was mainly subsistence oriented and that the main agricultural activity was growing ragi, almost all of which was for domestic consumption and distribution of wages. Commercial agriculture, although desired by the farmers, was only a distant dream under the prevailing farming conditions. There had been attempts in the past to raise a commercial crop now and then by the farmers. For example, during 1968-69, a Grama Sevak (village level worker, of the community development department) was successful in inducing three farmers to grow cotton. After one crop, these farmers gave it up because they failed to get a good crop and a good price for whatever they harvested. They also felt it to be a difficult proposition as the labourers in the village had no skill in many of the operations involved in raising cotton crop.

Farmers indicated their desire to become progressive and "rich" like the ones in the more irrigated regions of the state. This was possible through irrigation, which permited them to grow sugarcane and paddy. But, they face the problem of inadequate water supply as a result of which the farmers of Rajapura cannot take to , not only large-scale sugarcane or paddy cultivation, but also any other commercial crop. Wild bears inhabiting the forests in the region pose a threat to whatever little sugarcane and groundnuts grown. However, there are a few farmers who, from time to time grow sugarcane on cost sharing or sharecroping basis. This is mainly for the production of jaggery for domestic consumption rather than to sell either cane or jaggery in the market.

Given the limited underground water resources or rainfall, the introduction of sericulture was the best thing to happen to the farmers of Rajapura and to its economy. For, in the words of Charsley:

> It has the primary role, filled by a wide range of commercial crops elsewhere, of providing a major source of cash income to meet the cash requirements inevitable even for those who are producing their own staple foods, both for living generally and for production itself. In addition, in a drought-prone area, it provides a source of extra security since mulberry and worms can yield some cash income under almost any conditions (1982:97-8).

With the available water resources or in rainfed dry land farmers were now able to engage in an activity that brought them some monetary income. Sericulture was being described by the villagers as the blessing of Lakshmi—the goddess of wealth and prosperity. With sericulture, a farmer with 0.25 acre of rainfed land could raise at least three to four crops a year. Whereas, a farmer with an assured source of water could raise as many as six to seven crops a year. There were also some who raised more than seven crops by rearing worms on two different lots simultaneously with a difference of 10 to 15 days in egg hatching, thus rearing worms throughout the year.

Not only was sericulture a new source of income, it brought income more often in a year than in agriculture. On an average sericulture provides cash income at the end of every two months. It not only helped such registered farmers[6] engaged in full time sericulture, but also such others who did not actually take to sericulture or those who were not registered sericulturists.[7] Some farmers, for example, grew mulberry on their land only in order to sell the leaves to the farmers needing additional supply of leaves. During my study there were as many as twenty vacant house-sites which were being used as rainfed mulberry garden. Many Holeyas and Madigas had grown mulberry not only on the free house sites alloted to them, but also in their backyard or in front of their houses. A well grown "tree' of mulberry (*marakaddi*) was fetching Rs. 25 for a harvest of leaves.

While the labourers were paid both in kind and cash for agricultural work, they were being paid only in cash for all work connected with sericulture. There was also an increase of wages from 0.50 paise during the early '70s to Rs. 2.00 per day with food during 1978-80.[8] Annual payments to the services of the artisans were now being made in cash for the additional services rendered on account of sericulture.

For sometime payment for cocoons at the marketing centre was made through commercial banks, which compelled the farmers to open savings bank account. In order to induce the farmers to save money the officials issued part of the payments through National Savings Certificates, maturing at the end of six years. However, many farmers withdrew money from Saving Bank Account in order to make payment of wages to the labourers, and to people from whom they may have bought leaves or hired appliances. All such payments, including wages, are made normally after the cocoons are sold in the market and money received from the Department of Sericulture, or the Licensed Seed Preparers (LSPs). There were a few farmers with a bank balance of Rs 5,000 to 15,000 at one time or the other. Most farmers withdrew their money sooner or later to invest in land, gold or to build or repair their houses, buy cattle, or to spend on marriages, births, deaths, or illness in the family.

Sericultural Technology: An Introduction

Before examining the impact of introduction of sericulture, I shall briefly note here some salient features of sericulture and of some activities in which the farmers took part.[9] Such an introduction is necessary to appreciate the changes that took place in the light of sericulture. In many parts of Karnataka state, mulberry cultivation and rearing of silk worms were widespread.[10] In a sense sericulture is an agro-based industry, involving cultivation of mulberry and rearing of silk worms on the one hand, and reeling silk from cocoons on the other. Reeling may be taken up either as a cottage industry or in large scale factories. My discussion here refers mainly to the cultivation of mulberry and rearing of silk worms, and since I deal with a seed area, the details mentioned pertain to that.

As a first step in the cultivation of mulberry, irrigated or rain fed soil has to be prepared for planting the mulberry cuttings. This is generally done during July-August, so as to coincide with the monsoon season. The cuttings are usually bought from other farmers when they prune their plants. The cuttings are planted in pits of 35 × 35 × 35 cms. Compost or cattle manure is applied at the time of planting. With timely watering and manuring the plants reach a height of about 2 meters in about 6 to 8 months. Weeding is carried out at least twice during the first six months; once after 2 months of planting followed by another, with an interval of two or three months. In fact, weeding is done by digging the soil deeply around the young plants. The plants are to be pruned at the bottom every year. In the case of rain-fed mulberry gardens, it takes nearly 10 to 12 months before pruning. Before this, two small harvests may be carried out — during November/December and April/May. Pruning is made to keep the plant in a state of vigorous growth and to obtain good quality leaves. Leaves harvested are fed to the silk worms.

Rearing of silk worms has to be done in accordance with the guidelines provided by the department of sericulture. The important requirements are a separate rearing room, appliances like the baskets, trays, mounts, and a tray-stand, all of which are to be disinfected after every use. Most farmers use a portion of their house for rearing the worms. After a severe loss of crops during 1980-82 due to the attack of Uzi flies farmers have now begun to use a net to prevent such flies infesting the worms. The officials of the department visit the village frequently to disinfect the rearing houses and other appliances. Farmers have to further ensure that the quality of the leaves with which the worms are fed are good and uniform in quality. Leaves that are picked should be preserved in moist conditions. Magadi taluk has been identified as a 'seed area', where the cocoon production is exclusively used for seed purposes. Farmers, therefore, are to compulsorily sell their harvest of cocoons to the department's markets or through them to the LSPs. Further, they have to obtain eggs for rearing the worms only from the department's grainages. When assured of a supply of leaves, farmers apply for layings of silkworm eggs, which were sold at Rs 5.00 per 20 layings, each consisting of 300 to 500 eggs.

On receiving eggs, farmers have to wait for three to five days for the eggs to hatch into tiny worms. From now on the worms are to be carefully looked after by feeding them properly and regularly and storing them in trays. Silkworms are voracious eaters of mulberry leaves and grow rapidly. Commencing from the stage of *chawki*, they pass through five stages of growth. At the end of each stage, there is an interval of one or two days of rest for the worms, referred to as 'fever' when they do stop all their activities. On other days, farmers are extremely busy picking leaves, processing them, feeding the worms, and transferring the worms each morning from one tray to another, and in the process removing the waste from the trays. Such waste has excellent use as manure, as also fodder to cattle. The frequency and quantity of leaf consumption by the worms are more during the latter stages of growth. During this period farmers have to often keep awake during nights to feed the worms. The garden has to be watered regularly to maintain the requisite level of moisture in the plants and leaves.

When the cocoons are formed they are picked and taken to the cocoon market, at Doddi. In such markets the department has the exclusive right to buy cocoons for seeding purposes. If there is inadequate demand and if the quality of cocoons is good enough,. the department permits the farmers to sell the harvest to the LSPs. If there is no demand from the LSPs also, or if the quality of the harvest is bad, the farmer has to sell them through auction for reeling at Ramanagaram market. During 1978-79, being referred to the reeling market at Ramanagaram meant a financial disaster to the farmer. For, instead of getting something like Rs.40 and upwards per kg., the farmer got at best Rs 8 to 10 per kg. of cocoons. At times the cost of going to Ramanagaram was much more than the returns. Since 1982 a farmer is given a bonus of Rs 10 for selling at Ramanagaram, and the market is now regulated to prevent loss to the farmers. I learn that at times now it is profitable to sell cocoons at Ramanagaram than in Doddi, for, in addition to the bonus, the price offered for seeding is much lower than reeling at times. However, the weight of the cocoons determines the price: 1,000 cocoons weighing one kilogram got an average of Rs. 20 to 25 during my study. This was the minimum prescribed by the government, although a farmer may get more if the crop was good and if there was demand.

About 40 years ago also Rajapura had, but only a small number, sericulturists. The activity was more like a hobby than an economic one. It spread among others but mainly among those who had plenty of land and domestic help. Even among them the practice was by crude methods. Leaves were obtained from overgrown rain-fed mulberry plants. Many of the appliances that are now in use were then unknown and the quantity of worms reared was small. Instead of using worm-rearing trays or mounts, the early sericulturists used dried up branches of small trees or shoots of dried mustard plants. Naturally the need for other appliances such as mounts, a stand, or for a separate rearing house never arose. Some farmers informed that in the past (about 40-50 years ago) they collected the eggs from wild plants and trees in the forest.

In 1967, a grainage and marketing centre were opened in Magadi town. Prior to this, farmers had to go to far away places like Kunigal, Channapatna etc., to sell the small harvest, or wait for a visiting merchant to buy them.[11] The coming of a grainage at Magadi town gave a boost to sericulture in this region, but the farmers of Rajapura began to take up sericulture in large numbers following the opening of a marketing centre at Doddi village in 1969.

Changes as a Result of Sericulture

I shall now examine the impact of sericulture in the village. I shall begin by noting some of the changes in agrarian social structure, followed by noting a discussion of changes and continuities.

Pattern of Land Use

Since 1969, there was a gradual increase in the area under mulberry cultivation. As was noted earlier, farming activities in the village centered around cultivation of ragi and other crops grown along with it. In the past also mulberry was grown, if at all, in fences or here and there in fields with no separate plot of land earmarked for the purpose. Further, sericultural activities were restricted only to agriculturally slack seasons. With the establishment of the office of sericultural department at Doddi, things began to change. Since it was necessary to have a separate plot of land mulberry, farmers now began to allocate a piece to mulberry garden.

It is difficult, however, to grant credibility to the data on area under mulberry as recorded in the offices of the Revenue and Sericulture departments. Apart from the reasons inherent in the recording system, unreliability of data also stems from the fact that the farmers tended to give inflated account of area under mulberry so as to obtain more eggs during the peak season or to become eligible to receive various appliances from the Department of Sericulture at subsidised rates. The data in Table 4.2 has been compiled after cross checking it with farmers in the village. It is estimated that about 17 acres of land were under mulberry cultivation in the village during the early part of 1970 among the 44 early sericulturists. On the basis of 1971 census figure of 143 households, one may estimate that about 31 per cent of the households during the early 1970s had taken up sericulture. The increase in the number of households during 1971-78 was only by forty, while the number of sericulturists nearly tripled during the same period. The average size allocated for mulberry was 0.25 acres in case of rain-fed mulberry; 0.50 acre by sharecroppers; 0.50 to 2.00 acres by farmers with irrigated gardens. The increase in average area under mulberry per household has been only by 0.10 acres.

Irrigation

The second change as a result of sericulture is development of irrigation in the village. In an interesting study in a drought prone region, MacLachlan (1983) pointed out that one of the ways of surviving the drought conditions in the region was through irrigating the lands intensively (1983). Rajapura too which, in the past, was a dry village, could now boast of as many as 20 irrigation wells, while there had been one or two in the past. Over the years, the number of irrigated mulberry gardens has increased from 18.2 per cent during 1970 to 67.2 per cent during 1978-79. While in 1970 there were only 3 diesel/electricity run irrigation pumps, there were now 23 such pumps. With a lone exception, all the farmers had invested in irrigation pumps by borrowing from the land mortgage bank at Magadi. This finding goes contrary to the observation by Dasgupta (1977) in West Bengal. Two factors may be suggested as explanation for the pattern in Rajapura: (a) lack of savings, and unavailability of credit from informal sources; (b) attractive subsidies for loans.

Demand for Irrigated Land

As a result of sericulture there was an increased demand for land with irrigation facilities, either actual or potential. Not all land in the village had favourable underground water resources. The wet-belt of lands in the village began to assume an increased economic significance in matters of sale-purchase, and inheritance. This, however, does not imply that such lands did not have a special value prior to the coming of sericulture. Possession of wet land has always been every farmer's desire. But now with a cash crop farmers began to look for these lands, or convert dry land into wet, garden land. In Rajapura this can be seen by the fact that wells, as deep as 100 meters or more have been sunk in the dry belt. Yet, many have inadequate water during the summer months. Wells in the wet belt are generally shallow (20 to 30 meters), and water overflows from them during rainy season. The monetary value of land with an assured water supply rose as a result of increased demand for such land. It is rather difficult to assess the increase in value in absolute terms as there is hardly any sale of such land. In 1974, a tiny plot of 0.15 acres of irrigated land was sold for Rs 15,000, while the price of dry land was not more than Rs 1,000 per acre. During 1979-80, however, four acres of dry land around the wet-belt, with only a potential of irrigation, was sold at Rs.9,000 per acre.

TABLE 4.2. Area under Mulberry (1970-1980) (area in acres)

Year	Area under Mulberry											
			Irrigated						Share			
	Dry		Lift		I.P.set		Diesel		cropper		Total	
	No.	Area	No.	Area	No.	Area	No.	Area	No.	Area	No.	Area
1970*	42	13.63	5	1.50	3	1.50	—	—	NK@		44+	16.33
1980	40	12.34	36	14.9	21	20.49	3	1.00	20**	5.75	199	53.67

* Source for 1970 data has been mainly from the records of the department of sericultural office at Doddi; crosschecked with the farmers of the village.

@ Not known

\+ The total does not tally, because some farmers had simultaneously cultivated dry and irrigated land

** Includes only sharecroppers of irrigated land, and excludes tenants and sharecroppers of dry land or lift-irrigated land.

With the increase in the demand for irrigated land, the process of division of family property became a more complicated one than it normally was. All the inheritors now make an uniform claim for a lone well and the land around it. None of them wishes to take a risk by accepting a share of bigger dry land elsewhere and later invest in sinking a well. It is, therefore, not uncommon for mediators[12] to divide the property in such a way a well is set aside as a common property and the land surrounding it to be distributed among all the brothers. Water from the well is to be shared by all by mutual consent, just as the cost of upkeep of well, irrigation pump, etc. Such arrangements are not friction free, and sooner or later there will be disputes over rights to draw water, one alleging that water was diverted by another, or that one had not paid his share of the electricity cost, and so on.

There are also cases of farmers 'buying' water from neighbouring farmers if they do not have a well or adequate water in it. A farmer paid Rs. 20 to 30 per month for an hour's supply of water on alternative days from his neighbour's well. Sale of water is something new to the village, which was only after the advent of sericulture. Aged farmers speak of this as if it is an amoral activity.

Labour Specialization and Creation of New Roles

Another important change was the creation of a specialised class of labourers, such as the construction workers. There was now an increased demand for construction labourers for tasks like sinking wells, repairing the existing ones (e.g, stone layering), construction of rearing houses (or altering the existing ones), and building pump-houses. Creation of this class of labourers also meant withdrawing of such labourers from agricultural labour force, at least partially.

In the past, there were only crudely finished wells, referred to as *mannu bavi*, or *pillu bavi*, which used to dry up during the summer months. From 1969 onwards farmers began to sink proper wells and/or improve their wells by way of stone lining around the well. By 1966 the first power-run pump had been commissioned in the headman's fields. Many others began to emulate him by first sinking a well, and later installing an irrigation pump.

Hitherto, there had not been a separate category of construction workers as such, although there were some known for their skills who also worked as agricultural labourers. But, following the increased construction activities, there emerged a group of specialised labourers resulting in a considerable shortage of agricultural labourers. In addition to about eight households engaged in construction, there were two to three groups of migrant workers. The latter took up building a house or sinking a well on a contract.

Almost every farmer with a well undertakes to 'clean' his well annually, by way of desilting and deepening it. Every year, there are at least two to three farmers engaged in either sinking a new well or cleaning it. Construction of houses also became an important activity since rearing worms required spacious houses. In most cases, old houses were renovated or wherever possible, a new room was built. Renovation became necessary for a variety of factors: to protect the worms from smoke, insects, and stray animals; to prevent the entry of lower caste functionaries of the department of sericulture into the inner parts of the house where the worms may be kept; to have more light and movement of air in the rearing rooms, etc.

In addition to the creation of specialised labour groups, there was also development of new roles, namely, middlemen and informal technical experts in the field of sericulture. Farmers often face either a shortage or surplus of mulberry leaves in their garden. They may have to buy or sell leaves as the case may be. There are some transactions for which there is no standardised measurement either of quantity or quality, as for instance, manure, paddy or ragi saplings, or fodder. There is no fixed price for the leaves, nor is there a means of accurately measuring the quantity or quality. The prevalent criterion is to assess by looking at the basket containing leaves, or by looking at the number of leaf-bearing mulberry plants, referred to as *patta*. There is always a bargaining in fixing the price of leaves, which, in turn depends upon the demand for leaves in a given season, the quantity being bought or sold, and upon the nature of garden in terms of irrigated or rain-fed.

In order to have a smooth transaction, a middleman's services are obtained. Such middlemen generally enjoy the reputation of being good judges of leaf quality and for being fair to both the parties concerned. The buyer and seller ask the middleman to inspect the garden and to suggest a price for leaves to suit both the parties. There may be a further bargain in either lowering or raising the price. The middleman may receive a nominal sum of Rs. 5.00 per transaction. There are also instances where the middleman rendered his services without accepting any fee. Whether paid for or rendered as an honorary service, the need for middleman is important since often the parties involved in a transaction are close relatives or friends, or even bitter enemies. The presence and role of a middleman paves the way for a smooth transaction, and the value of the goods determined in an objective way.

Such middlemen have to be fair minded with no bias towards either of the parties, for their reputation in future will be at stake. The concern for personal reputation was seen to be lessened when the role of middlemen was in a sense institutionalised. This came about in 1978, with the establishment of a *chawki* centre. The centre was opened with the intention of rearing young worms in a scientific way, since the end result was dependent upon the way *chawki* was reared. When inaugurated, the centre had rented a house (Rs. 200 per month, the highest ever by local standards), and equipped it with stands, worm rearing trays, and the other requirements. It was attended by two daily wage labourers from Rajapura. A committee was constituted to look after the affairs of the centre: hiring out the appliances, auditing the income and expenses, and to buy leaves from farmers for the centre. A farmer needing the services of the *chawki* centre had to deposit the silkworm eggs with the attendants, and pay for the cost of leaves and a nominal service charges. It was alleged that the members of the committee usually selected their own or their supporters, mulberry garden for purchase of leaves, and that they determined the price arbitrarily with no regard for quantity and quality of the leaves. It is important to note that some of them were also known to be middlemen, but in their capacity as members of the committee they did not care much for their reputation. In due course, however, the practice of the centre buying leaves to feed the worms was given up. Instead, the farmers were asked to supply the leaves themselves along with the eggs. This was also because there were frequent complaints about the quality of leaves used in feeding, causing loss of crops.

In addition to the role of middlemen, there were also informal technical specialists who were now consulted by the farmers in all matters related to sericulture. These technical experts gained their reputation as being experienced in all matters of sericulture not only by their past performance but also for having won commendation from the department for participating in workshops, exhibitions or taking out record harvests. Some of these people have taken to sericulture as their full time occupation, and agriculture has become a subsidiary one. Rarely have these experts faced a loss of crop, and normally got a good price for their harvest. They are always consulted by almost all farmers in Rajapura right from the time when they convert a plot of land into mulberry garden, and in matters of rearing worms. Their opinion is sought about the growth of worms, as to what steps are to be taken to improve their growth, what kind of leaf is to be fed, or what to do when worms show signs of diseases. Often they recommend folk methods of overcoming the problems, and their opinions clash with those of the employees of the department.

Of course, these experts gain no monetary benefit by advising the farmers, but enjoy the social status of being important men in sericulture. They do get some indirect benefits, sooner or later, such as a plot of land for tenancy cultivation or ragi from a regular 'client', as one expert Ganganna once received free wood to build a house. None of the recognised informal experts were, however, nominated to the committee to supervise the *chawki* centre, although the attendants consulted them.

There were also another set of roles emerging in the context of sericulture, namely, the agents of the Licensed Seed Preparers. The LSPs purchase cocoons from the farmers through the sericultural markets. As the LSPs cannot be physically present in all the marketing centres always, they engage the services of paid agents. These agents are authorised to purchase cocoons on behalf of the LSPs from the farmers. The agents receive a commission of 10 per cent of the proceedings, which, however, may vary depending upon the quality of, and demand for cocoons.

These agents also indulged in scouting a good crop of cocoons in the surrounding villages. Having spotted a good crop that is likely to reach the market in due course, the agent or the LSPs often came to an understanding with the officials of the market to permit to buy such a crop. At times the agents also persuade the farmers to bring their harvest to the market on a specified day which is convenient to the LSPs or their agents to be present there. In Rajapura there were five such agents, who from time to time acted on behalf of the LSPs. There were at least ten such LSPs and many more agents who operated at the marketing centre at Doddi.

Before concluding the discussion of development of new roles, mention should also be made of different kinds of participation of people in sericulture. Having seen the benefits of sericulture, many people now wanted to be engaged in it although they did not own any land, or land suitable to grow mulberry. Some of them were not even registered with the department of sericulture. During the course of my study, there were as many as ten such 'sericulturists' engaged in rearing worms throughout the year with the leaves bought from others. They had procured the eggs through one of their kinsmen or fellow farmers, and sold the harvest of cocoons in their names. As regards appliances, they may have borrowed or hired from others or used the ones which they may have had.

An interesting feature of such part-time sericulturists was that often to meet a forthcoming expense, a landless labourer could engage in rearing one or two 'crops' of leaves and worms. For instance, an unemployed youth, or a landless may approach a landowner who is not currently engaged in rearing worms[13] to allow him to make use of the appliances and the leaves in his garden for a crop or two so as to enable him to meet a particular financial need. Once the harvest is sold in the market or to the LSPs, the part-time sericulturists make payment for the leaves and labour they may have hired. Usually, in such arrangements the landowner may not collect any rent for the use of appliances like a wooden stand, trays, or baskets. However, rent has to be paid for hiring mounts. The extent of dependence of such part-time sericulturists upon others can be seen from the fact that, apart from borrowing the wooden stand, trays, and hiring of the mounts, they also often seek rent-free space from others to rear worms. Through such arrangements a person is able to earn at least Rs. 400 or more by rearing worms from 100 layings at a time.

There are also farmers who regularly sell leaves from their garden, without rearing worms themselves. Some of them procured eggs and sold leaves to both registered and unregistered sericulturists. One such farmer had a power run irrigated pump installed for his half acre mulberry garden. The reasons for some of them not rearing worms by themselves were many: other lucrative activities like trading in cattle fairs and weekly markets; not enough members to support sericulture in the household; undertaking civil contract work with the Public Works Department and the then Taluk Development Board; interest in local politics which took them away to town almost every day, and so on.

Increased Demand for the Services of Artisans

Sericulture requires, on the agricultural side, preparation of soil crop after crop, and bottom pruning of mulberry plants. This implied an increased use of the tools and implements as well as making them anew and repairing them. In connection with worm rearing, the farmer require trays, tray-stand, mounts, baskets for carrying leaves from the garden.

As a result, the blacksmiths, basket-makers and the carpenters in the village were found to be constantly in demand for their services. Prior to the advent of sericulture the local smiths were busy only during certain months in a year. The demand for the services of the basket makers and blacksmiths has been steadily increasing over the years. There were three basket-makers of the Korama caste. In the past there was a suitable environment, as plenty of bamboo grooves were available not only in the nearby forests, but also in private holdings. Though it is now forbidden to cut bamboo trees in the forest, the basket-makers have been managing to get enough supply of bamboo with or without the 'permission' of the forest officials. Farmers also sell bamboo grown in their land. The general practice for a basket maker is to contract with farmers to fell bamboo trees and make about ten bamboo trays, and a few baskets in return for the bamboo. He may also make these for a wage payment, or make on his own and sell them later.

On an average, a farmer with 0.25 acre of mulberry garden requires at least 20 worm rearing trays, forty mounts, and about five to six baskets for carrying leaves. Although most farmers may not possess them in requisite numbers, from time to time they need to repair those which they have. In addition to the demand for the services of basket-makers for sericultural needs, they are also required to make other items like bamboo knitted screens, and winnowing trays etc. As a result the basket-makers are in demand throughout the year. Prior to 1970s they used to visit weekly markets in Magadi or elsewhere to sell their ware, whereas they are now waited upon by the farmers at the former's work place. They cater to the farmers from neighbouring villages also.

In addition to meeting the requirements of the farmers, the basketmakers are also busy for another reason. The department of sericulture undertakes to supply trays and mounts made up of bamboo at subsidised rates to the small and marginal farmers as well as to the members of the Scheduled Castes and Scheduled Tribes. This has led to another specialised role, namely a contractor. The contractor undertakes to supply the appliances, and hires the services of local basket-makers to make them on a large scale. One such successful contractor was from Rajapura, who is a brother of the former *Patel*, and a close relative of the headman. These contractors keep the basket makers busy throughout the year, who find it easier to ply their trade this way than having to obtain bamboo on their own. Rajapura's contractor was so successful that he earned a profit of Rs. 35,000 in

1979, and had obtained another major order to supply in the neighbouring taluks also. Since the local basket-makers did not have sufficient manpower and raw materials the contractor went to another taluka to get the appliances made in the subsequent years. To indicate the extent of success of this contractor, I may mention here that in 1980 he bought a plot of five acres of land in the outskirts of Bangalore city, which he later sold as house-sites. The same year he also concluded a deal to buy 5 acres of land of the Lingayat landowner in Rajapura, at the rate of Rs. 8,500 per acre.

There has been substantial gains for the basketmakers also in the process. One of them, who was landless hitherto, bought two acres of land, and sunk a well and installed an irrigation pump. Consequently, he has also been shifting his attention more to farming than to basket-making. One of his sons is now a Doctoral student in history in one of the universities in Karnataka. The other basket-makers have now more or less given up working as agricultural labourers and seasonally migrating to other places. They all feel that the department of sericulture would do well by avoiding the contractors who supply the appliances and assign the task directly to the craftsmen. Referring to a scheme to develop rural artisans, one basketmaker pointed out that the government does not think of proper means of improving their lot but only wants to advance loans. He said that he would better off if craftsmen were given work and be paid for it than their working for others who take the bigger share of returns.

The other craftsmen whose services were increasingly in demand were the carpenters. Rajapura has two carpenters, one of whom is an immigrant smith and also an experienced carpenter. The other is an Okkaliga landowner who learnt the craft from the immigrant smith. In addition to these two, a few migrant carpenters undertake assignments and camp in the village. The carpenters are generally commissioned on a contract basis, to make wooden stands required in sericulture, to make doors and for other requirements for constructing houses. With the increased prosperity through sericulture farmers also get furniture made by the carpenters. Rajapura's carpenters spend more time working in other villages. This is because almost all farmers now have their wooden stands made, or they do not require much upkeep of the older ones. The smith-cum-carpenter and his two sons have an establishment in Bangalore city, while another son lives in Rajapura.

Finally, the blacksmiths too have now an increased demand for their services. The two blacksmiths have had a history of frequent migration to other places in search of employment in the past. In contrast, they now engage in repairing or making tools and implements throughout the year. Although they have been spending their income in unproductive ways, they are now better off than ever before. In addition to several other service specialists in the village, the blacksmiths have retained their role in the network of *jajmani* relations. Notwithstanding the coming of a commercial crop in the region the institution is persisting (for a detailed account of this, see Karanth 1987).

Increased Interaction with the Neighbouring Villages

In more than one way Rajapura's relations with its neighbouring villages increased both in terms of frequency and purpose following the advent of sericulture. Doddi, with its sericultural offices, became an important centre for the people of Rajapura.[14] It also houses the credit cooperative society, Panchayat and village accountant's office, rural health centre, veterinary clinic, and schools. Farmers from Rajapura visit Doddi more often to book their requirement of silkworm eggs, market the harvest, to hold consultations with the officials. Since there are more number of buses plying through Doddi to either Magadi or Ramanagaram, Rajapura farmers go there to take the bus than walking to Magadi as they did in the past. In a sense the increased monetary income has also contributed to this.

In turn Rajapura became an important place for its neighbouring villages as it had the largest number of sericulturists under the Doddi range office. Farmers from other villages now began to visit Rajapura in order to borrow or hire sericultural appliances, or to purchase leaves, cuttings of mulberry plants, and to consult the folk experts. Relatives living in Rajapura and its immediately adjacent villages used to have exchange of goods and services. For example, a farmer borrows trays and mounts that are additionally required from a relative living in any of these villages on a reciprocal basis in order to avoid paying rent for such appliances. At the peak of a crop (when the worms begin to form cocoons) such relatives and friends visit each other to help, which is reciprocated. Such mutual help often cuts across the village boundaries. Similarly, farmers along with an advising middleman from other villages visit the gardens in Rajapura to buy leaves. To cater to the 'floating population' there are new tea-shops and grocery stores. After the inauguration of the *chawki* centre people from all the neighbouring villages are coming more frequently to place orders for *chawki* and to hire mounts from the centre.

Increased Participation of Female Labour

Traditionally women have had some particular roles to play in agriculture. A woman in a typical household engaged herself in assisting the men in farming activities: for example, at the time of ploughing and sowing she may visit the fields either to join or supervise the work of other female labourers in picking weeds, assist her husband or son by carrying seeds and inserting them into the *kurige* (an instrument for sowing ragi and other seeds); carry food to the men and other labourers working in the fields. During the season of *avarekai* and groundnuts she may pick the produce. In addition, she also manages the grain stock at home, and the cattle. However, the nature and extent of women's involvement in agriculture depended upon caste and class to which the household belonged, as also upon other factors like the number of men of working age or the size of the household, whether or not they also worked as wage labourers, the demand and supply of labourers during a given season.

It was not uncommon in the past for adverse comments (such as miser, greedy, etc.) to be passed if women of a landowning household worked in their fields notwithstanding their economic status. With sericulture now, women are drawn into physical labour more than ever before and without any social stigma attached. Women take part either independently or with other hired labourers in picking leaves and in removing the waste from worm rearing trays, feeding and changing worms from tray to tray, cleaning and chopping the mulberry leaves, and in picking cocoons. Women also assist men in regulating the water flow from the pump, and diverting the flow from row to row in the mulberry garden.

There is now a tendency for female labourers to specialise in only sericultural operations while rejecting work in any other agricultural operations. The exception is, however, during weeding and harvest seasons when they work independently or in exchange labour groups. Since their menfolk also work for wages in other operations and are paid accordingly, women labourers collect their wages usually after the harvest of cocoons. It is also not uncommon for them to deposit their earnings for safe keeping with the employing landowner, or indulge in borrowing foodgrain and money from time to time. Most female labourers have now come to establish a more or less enduring relationship with particular landowners in the sense that they work only for such and such a landowner throughout the year, crop after crop. Such women are also employed for other agricultural operations when there is no sericultural activity.

Cultural notions of ritual purity and pollution have found place in sericultural operations involving women. Not only do farmers refuse the services of women during their menstruation, but also the latter refrain from working on such days. It is believed that their coming in contact with the worms adversely affects the growth, and even results in the total failure of a crop. An extension of this notion is now in the form of a taboo for such women to enter a mulberry garden also. However, such notions of purity and pollution apply only to the hired women labourers, and not to the members of landowner's household.

Sericultural folk experts offer the explanation for such beliefs on what they call a scientific ground. It is claimed that women have an abnormal body temperature during the menses period and it affects the health of the worms. The officials in the sericultural department, however, ruled out any association between the two. From the point of view of people, it may be pointed out women are considered to be ritually impure during their periods. Since farmers attach a special value to sericulture, as goddess Lakshmi who has come to the rescue of the dry-land farmers, they apply all notions of ritual purity and pollution also in the field of sericulture. Similar notions of purity and pollution prevail in agriculture in general also, particularly during the harvest operations.

So far I have described some of the important changes that have taken place as a result of sericulture. I shall now proceed to examine some of the ways in which persistence of the old or 'traditional' patterns of agrarian relations was witnessed following the introduction of sericulture. Some such continuities could be witnessed even within the scope of changes noted above, as for example, continuity of *jajmani* relations with the smiths and other service specialists despite a commercial crop in the village, interdependence of farmers and labourers, enduring relationship between female labourers and landowners, etc. In order to account for these and other continuities I shall first examine the limitations of large scale sericulture. My contention is that it is because there are certain limitations to intensive and large scale undertaking of sericulture that some of these continuities occur.

Constraints to Large-scale Sericulture

Although coming of sericulture is considered as a boon to dry-land farmers, it could not be easily taken up on a large scale. Some of the major constraints are shortage of underground water resources and labour supply, variations in the price of cocoons, and frequent failure of crops due to various diseases of worms. Most wells in Rajapura are on the verge of drying up during summer months when the price of cocoons in the market is generally higher. As noted earlier, not all tracts of land have a favourable water resources. In such tracts with relatively plenty of water, the land has already been highly fragmented. Investing in sinking wells, pumps etc., in such small holdings may not be economically feasible. Furthermore, even if a farmer has a fairly large holding with an assured water supply for a mulberry garden, of say two acres, he is faced with the problem of finding labourers regularly for all sericultural operations. He has to have a fairly large household with at least two or more male members to assist him in sericultural as well as general agricultural operations. Farmers find it difficult to obtain labourers in time, and further find it difficult to totally depend on either domestic or hired labourers.

Over the years there has been a gradually increasing scarcity of labourers. This is chiefly because of sericulture itself. There are many labourers who have now become sharecroppers in sericulture, hence, restrict availability of their service only to a few farmers with whom they sharecrop. In addition, the creation of specialised labour, as in construction, middlemen, etc. has also affected labour availability in both sericulture and agriculture. This shortage appears to be a more recent development, for, it is said that about forty years ago labourers used to wait near a temple yard each day for prospective employers to hire them. Many a bonded labourer in the past had entered such a service only to ensure food and work, although a need for a loan was also a factor. Further, there are many labourers who had migrated and were in the habit of seasonal migration in search of work. Another aspect of shortage of labour is that there is no separate category of labour households as such, with an exception of a few non-Okkaliga caste households. A bulk of the village's labour supply comes from the marginal and small farmer households, while most landless households now have sharecropping and tenancy arrangements to supplement their income. Almost all landless households also engage themselves in gathering and selling firewood, grazing and rearing cattle, and construction work. Some work as hired labourers to lay trenches in the forest and in nurseries, or to collect tamarind for the contractors who lease the trees in the forest. For such assignments they are paid a higher wage and can work under far more liberal working conditions than in the agricultural operations. Further, labourers are lured with soft loans to work for the contractors in repairing roads, tank bunds, etc.

As a result of all these factors, there is now a short supply of labourers. This comes in the way of undertaking large-scale and intensive sericulture by a farmer who may have adequate water resources and land. Thus, although sericulture is a far more profitable source of income, the farmers of Rajapura face certain constraints like a scarcity of suitable land, labour, and water resources to undertake it on a much larger scale than possible under the circumstances.

In addition to these constraints, there is the insecurity and uncertainties involved in sericulture, which are, however, not peculiar to sericulture alone. Silk worms are extremely sensitive creatures, being vulnerable to changes in climate and to a variety of diseases. Threat to their safety and health comes from ants, flies, lizards and stray animals. Farmers face many obstacles in providing more hygienic conditions to combat these obstacles. Not all of them can afford to have a separate rearing house, nor can they have a smoke-free kitchen. The fear of loss of a crop, particularly for farmers rearing large number of worms, comes in the way of large-scale sericulture.

Further, farmers face another important constraint in terms of fluctuations in the market price of cocoons. As Charsley observes, "Having cocoons to sell one week, even one day, may mean wealth by local standards, whereas selling even the next day may mean achieving little more than a covering of costs or not even that" (1982:96). It is not uncommon that it varies from Rs. 70 per kg. on a day to Rs. 40 in the next few days. Cocoons cannot be hoarded or stored for long in order to wait for a favourable price. Even if the market price is good at times, the department may not have a demand for it on the days a farmer has a crop to sell. Consequently, he may have to sell it to the LSPs or to reeling in Ramanagaram. As a result the farmer is compelled to keep and rear smaller number of worms, and manipulate the cycle of rearing worms in such a way as to have a harvest at least once in a month or fifteen days.

The constraints discussed above, pose a dilemma to the farmer: the need for a good and frequent monetary income on the one hand,

and risks and constraints of large scale sericulture, on the other. Bigger landowners, particularly the ones with adequate water supply and irrigation pumps face this dilemma more than the others.

Continuity of Traditional Agrarian Relations

Farmers of Rajapura have found an answer to the dilemma pointed out. Their answer, in a way, has resulted in what may be broadly considered as continuity or revival of some aspects of pre-land reform agrarian relations. This revival occurred not only as an answer to the dilemma faced in sericulture, but also in the field of agriculture in general. These developments also served as a reinforcement of caste, and kinship loyalties, of the kind observed during the process of immigration in the face of plenty of land available for tenancy and sharecropping before 1960s. These loyalties, it may be recalled, operated to obtain land for relatives, castemen, and political henchmen, on purchase or tenancy occupation of land. Further, the continuity was also in a way seen to be the farmers' manner of assuring a "labour bank", i.e., a source of regular supply of labour. The most important of these continuities was leasing out land on tenancy and sharecropping basis.

Farmers who go in for intensive and large scale sericulture are generally those small and marginal farmers who convert all their land into mulberry garden and rear silkworms throughout the year; and also the bigger landowners who generally require a continuous supply of labour. The former type of sericulturists, sooner or later, feel the need for "growing one's own requirement of ragi".[15] Having converted all their holding into a mulberry garden, marginal and small farmers have to now wait till they save sufficient money to buy a piece of land. If and when land comes for sale, they will have to compete with others and rely upon the patronage of the bigger and influential landowners to succeed in buying. Failing to buy land, the small farmers seek a plot for tenancy cultivation of ragi or any other foodgrain.

The bigger farmers, on the other hand, have now earmarked smaller plots of land, converted into mulberry gardens, to let out on sharecropping or on fixed rent basis. The latter is usually for rain-fed mulberry gardens, while the former is practised in irrigated gardens. Most farmers with irrigated land have more than one sharecropper, while the farmer himself may retain a bigger plot for his personal cultivation. By leasing out on a fixed rent basis the farmer gets a monetary income without having to work on the land himself. By the end of a few years he may have a well maintained mulberry garden, for which the tenant would have worked. He may, however, have a separate irrigated garden also. An acre of rain-fed mulberry garden was being leased out for annual rent of Rs. 300 to 400 during 1978-79.

In the case of sharecropping, a farmer usually takes two sharecroppers to cultivate the land to get greater and more frequent income than what he may have got alone. In addition to this, the farmer gets an assured labour supply from the household of his sharecroppers, who, out of a sense of obligation, accept to work for him whenever commissioned. Such obligations are not restricted to the working on sericultural operations alone, but also for all agricultural operations. Such labourers do receive wages just as any other labourer does. It is, however, not uncommon for a landowner to seek wage-free services from them. For example, a sharecropper is asked to assist in necessary preparations for harvest, in putting up a haystack after ragi harvest, in making arrangements for a wedding, and so on. Just as harvest operations approach, every farmer in the village engages himself in making ropes, brooms, winnowing trays, and so on, required for transporting hay and grain. It is a common practice for bigger landowners to be assisted by their 'regular wage labourers', former bonded labourers, and sharecroppers. They are not paid any wages for this task although they may take home some of these items as may be needed by them. This type of landowner-labourer, tenant/sharecropper relationship existed even in the past. Its continuity in the context of a new cash crop goes contrary to the widely held view that commercialisation of agriculture has eroded traditional interdependence between the two classes (see, for example, Brown 1970; Singh and Singh 1975; Frankel 1972). The present interdependence stems from both their primordial and economic needs as may have been in the past. Most often, the sharecropper works on the landowner's personal garden as well while working on the plot on which he is sharecropping: fencing, making water-paths, watering the garden, and picking leaves. Landowner need not pay wages for such work since work is simultaneously done on landowner's and sharecropper's gardens which are generally adjacent. Some landowners also insist on the sharecroppers accompanying them whenever they go to the fields, or to the market in Magadi or to cattle fairs, etc. What is to be noted in this context is the nature of relationship between the two is very much like the patron-client relationship. It is not always that sharecroppers accept these obligations meekly. But resistance to heed to such demands may result in disciplining of the sharecroppers. Some minor forms of 'disciplining' a sharecropper is by discriminating against such 'erring' sharecroppers, as by not giving silk worm eggs in time or asking them to sell the leaves rather than allowing to use them for rearing worms, not operating the pump when water is required or inadequate, etc. Ultimately, the arrangement may be terminated, and another may be taken on as a replacement. There are always others eager to find sharecropping arrangements. One sharecropper remarked, after such an arrangement was terminated, that he was fed up of being a shadow of his landowner, for he was not only made to go with the landowner to Magadi or elsewhere, but also carry baggages and run other errands.

The terms and conditions for sharecropping in sericulture are that the landowner provides mulberry saplings (if the land is being newly converted into a garden),[16] land, water, and in some cases appliances for rearing worms. The sharecropper provides labour, and space for rearing worms. The cost of additional leaves, fertilizer, and rent for hiring additional appliances, if used, are shared by the two. If the sharecropper needs more labourers than available within the household, he meets the cost of wages. The sharecropper is responsible for the upkeep of the garden by weeding, fencing, digging the soil, and tending the plants and worms. He has rights neither over improvements made in the garden nor over the mulberry cuttings for his personal use when the plants are annually pruned. He does not have any rights also over any other crop for whose growth he may have contributed directly or indirectly (such as vegetables, coconut and other fruit bearing trees). However, these terms and conditions vary depending upon the nature of personal relationship between the two. For example, in some cases, the cost of hiring appliances may not be shared, or the landowner may meet all the costs. Concessions are usually shown if the sharecropper pleases the landowner by not only bringing out good harvests of cocoons, but also by accepting to work with or without wages whenever commissioned.

The practice of sharecropping and tenancy is found not only in sericulture, but also in agriculture. It may be recalled that as a result of land reforms, tenancy and sharecropping were on the decline. Their revival now may be explained with the same set of factors discussed in the context of their emergence in sericulture, such as the need for going into large scale sericulture, need for assured labour supply, etc.

Moreover, most farmers now pay greater attention to sericulture, and therefore, find it easier to let out part of their agricultural holdings for sharecropping and/or tenancy. In their present form, both sharecropping and tenancy usually refer to the cultivation of land for a specific crop, and for a specific period. The rent for tenancy cultivation is payable in kind or cash, after the harvest. The tenant provides the labour, seed, and all other inputs while the landowner provides land and pays the land revenue. Sharecropping arrangement, known as *palu*, requires the cultivator to equally share the produce with the landowner, while providing labour, and sharing the cost of manure, and seeds.

The general practice of sharecropping and tenancy to a given crop has an exception. There are about a dozen non-resident landowners who are in trade or service in other towns and cities. They hold a share in the family property in the village and usually lease out the land. The tenants and sharecroppers may take out any number of crops of their choice although the landowner has a say in it. Usually, these tenants are related to the landowner, if not as brothers. Whether or not a person cultivates the land of his non-resident brother depends upon the nature of division of family property and their interpersonal relations. In all cases of division of property there is an element of dispute, such as over matters like a claim for an additional share by the elder brother (*jeshtamsha*) with whom the old parents reside, propriety of the working brother demanding equal share of the property, and so on. The last mentioned is bound to occur if he makes an exclusive claim for such lands/or other property bought with his earnings from his urban employment. Usually, the other brothers who may be agriculturists, refuse to give an equal share to the salaried, and therefore, economically secure, brother. In all such disputed division of family property, the non-resident brother is unlikely to entrust cultivation of his share of land to any of his brothers. The point to be noted, however, is that social mobility through education and urban employment to the rural people is also likely to lead to continuity of traditional arrangements of cultivation such as tenancy and sharecropping. For, the rural educated and urban employed persons with farming background, in most cases, make an effort to retain their landed interests back in the village.

The distinction made between tenancy/sharecropping on land and for crop in the present day arrangements may appear to be subtle. But such a distinction has two important ramifications. First, an emphasis on crop precludes any possible claim by the tenants/ sharecroppers over the land under land reforms legislation. The land given to tenant/sharecropper to raise a crop is generally adjacent to the land personally cultivated by the landowner. This, and the emphasis on crop prevents any attempt on the part of the tenant/sharecropper to claim rights of possession. The second ramification may be explained in terms of 'formal' and 'informal' nature of the present day arrangements. The terms 'formal' and 'informal' here are not used to refer to any documents to be signed or legal formalities to be fulfilled. On the contrary, they are all oral arrangements. The formality or informality refer to the rigidity or flexibility in the terms and conditions of the arrangement.

Each farmer has a need for certain crops and their produce for household consumption. These are, for example, groundnuts, black-til, horsegram, jowar, and less frequently, sugarcane. Although farmers have a need for these crops, all may not be able to grow them independently on their own land for several reasons: inability to crop in time owing to preoccupation with other work; inadequate or unsuitable land; non-availability of labourers; ill health; etc. Therefore, a landowner frequently enters into an arrangement with another for cultivating the latter's land to grow a specific crop. It is here that the formality or informality of arrangements can be seen in operation. Conventionally, it is understood that such a 'tenant' shares the produce or pays a rent either in kind or cash to the landowner. When this is not specified, or not followed, a tenancy or sharecropping assumes informal nature. Informal arrangement does not involve any bargain or even a mention of the payment in kind or cash, or of the produce to be shared. The tenant casually requests for a plot of land (referred to as *gundlu*) to raise a specific crop. When the landowner begins to make arrangements for soil preparation, the tenant may also begin working on the specified *gundlu*. The tenant usually works on all the land and not merely on the *gundlu* while he shares only the produce from it. Such an arrangement, at the same time, assumes the nature of sharecropping/tenancy and mutual exchange of labour.[17] Implicitly in the present day sharecropping and tenancy arrangements, the tenant's vulnerability becomes a source of labour to the landowner. On the other hand, an arrangement may assume a formal nature, if the tenant does not work for the landowner, but only works on his *gundlu*, and shares the produce or pays a rent.

Irrespective of the formal or informal nature of the arrangement, the tenant/sharecropper usually despatches a share of the produce to the landowner's residence. This is because, a tenant does not like to have a reputation of cultivating a fellow villager's land without making a payment in return. Informal arrangements are generally found among farmers of equal socio-economic standing, often among kinsmen or fellow caste members, and among those with a well-knit interpersonal relationship. A tenancy arrangement between persons of unequal social and economic status tends to be formal, rigid and more exacting in the fixation of terms and conditions. Further, the nature of the arrangements (i.e, formal or informal) is also determined by the manner in which the produce is meant to be used. If a tenant takes up a plot of land for growing a crop in order to sell the produce, as in the case of mulberry, the arrangement tends to be formal, and often involves bargain in fixing the rent or quantity to be shared.

The principle of reciprocity of relations in all informal arrangements aids the landowner in more than one way: support and help in all his personal affairs or in matters of factional dispute. A person of unequal social and economic standing, on the other hand, may not be able to provide such returns in as many spheres of social, economic and political life as by a person of equal standing. A landless labourer or a marginal farmer can, at best, provide labour, which a bigger landowner can get for his 'labour bank' through formal arrangements.

In Table 4.3, I have classified sharecropping and tenancy arrangements into three categories: sericultural, agricultural, and both sericultural and agricultural. The data refer to the persons leasing in

TABLE 4.3 : Landownership and Incidence of Sharecropping and Tenancy

Size Class (in acres)	Share Cropping and Tenancy for							
	Sericultural only		Agricultural only		Sericultural and Agricultural		Total	
	No.	%	No.	%	No.	%	No.	%
Landless	4	15.4	8	30.7	14	53.9	26	100.0
0.01 to 2.50	2	11.8	10	58.8	5	29.4	17	100.0
2.51 to 5.00	Nil		10	55.6	8	44.4	18	100.0
5.01 to 10.00	Nil		10	90.9	1	9.1	11	100.0
10.01 to 20.00	Nil		4	80.0	1	20.0	5	100.0
20.01 and above	Nil		Nil		Nil		Nil	
Total	6	7.79	42	54.54	29	37.67	77	100.0

other's land, and not to the landowners who lease out. Landless and marginal farmers (with 0.01 to 2.5 acres) constitute nearly 69 per cent of all the households engaged in tenancy and sharecropping. Given the fact that a majority of agricultural labourers come from these two groups, it strengthens my contention stated earlier that the present day tenancy and sharecropping are offered by landowners mainly in order to ensure a 'labour bank'. Only the marginal and small farmers are engaged in sericulture (15.4 and 11.8 per cent respectively). Correspondingly, none of the bigger farmers with more than 2.5 acres was engaged in sericultural sharecropping/tenancy, although they were found to be engaged in the second and third types: 'sericultural', and 'sericultural and agricultural'. By reclassifying the three categories into 'sericultural' and 'agricultural' sharecropping and tenancy (the latter combining the second and third type), one finds that cultivation of other's land is more for agricultural purposes than sericultural, although it was sericulture which gave the impetus for such arrangements in the first place. It was pointed out in the discussion on land reforms that the practice of sharecropping and tenancy had declined as a result of the legislations. However, this practice had emerged later in the context of sericulture, which, in turn was in response to certain constraints to large scale sericulture. The data demonstrates that the revival of the practice has now gradually increased in agriculture than in sericulture. The data also suggests continued interdependence among different agrarian categories for their respective requirement of grain, labour, kinship and caste loyalty, etc.

Some Concluding Remarks

In this chapter, I have attempted to show how sharecropping and tenancy relations, which had declined in the wake of land reforms, revived as a result of a new cash crop. My purpose was to demonstrate the dual process of change and continuity in agrarian relations as a result of a change in cropping pattern. In this direction, I first described the village economy as it existed prior to the advent of sericulture, which was predominantly subsistence oriented. The main agricultural activities were centered around production of ragi. Lack of skilled labourers, inadequate water supply, lack of proper marketing facilities, the risk involved, etc., had discouraged many enterprising landowners from undertaking commercial agriculture. Sericulture was the only agro-industrial activity which promised the dry land farmers a cash income under the prevailing conditions.

The coming of sericulture had certain impact on the village's agrarian social structure: frequent monetary income to farmers and labourers in the form of wages; a gradual shift in importance in favour of mulberry cultivation and thereby bringing more land under irrigation. Yet another impact was an increased demand for land with relatively better underground water resources. Consequently, more and more irrigation pumpsets were installed. The increased need for irrigation and the need for construction of silk worm rearing houses gave room for the development of a new class of labourers specialising in construction activities. In addition to these changes, certain new roles began to develop in the context of sericulture: folk technical experts, contractors supplying sericultural appliances, middlemen and the agents of Licensed Seed Prepare, increased interaction with the neighbouring villages and towns. Rural craftsmen, particularly blacksmiths, carpenters and basket-makers, found an increased demand for their skill and services, what is more, under the 'traditional' *jajmani* network of relations. There was also an increased participation of women in labour force.

Along with these changes, the coming of sericulture also was responsible for a revival or continuity of some aspects of the earlier agrarian relations. Most important among them was the revival of sharecropping and tenancy, not only in sericulture, but also in agriculture in general. This revival was associated with the need to take up large scale sericulture in the face of inadequate water resources, shortage of labour, and the fluctuating market price for cocoons. The chief motive for the landowners entering into such arrangements was to ensure a regular labour supply from the households of tenants and sharecroppers. However, the present day sharecropping and tenancy arrangements were less enduring — on a year to year, renewable basis —, oral rather than being contractual, and more importantly, with a specific reference to a crop than to land.

Contrary to the widely held view, the coming of a commercial crop had not resulted in the erosion of traditional interdependence among different agrarian categories, nor had it resulted in any significant conflict or tension amongst them. This was substantiated in terms of loyalty of labourers towards the landowners, wage-free services, informal tenancies, etc. Interdependence was also demonstrated with reference to the 'registered' sericulturists helping the 'unregistered' in raising a crop or two of mulberry and cocoons to meet any contingency. I shall note more such evidence of continuities and also of changes in traditional agrarian relations in my discussion of landowner-labour relations in the following chapters.

Notes

1. See Table 2.2, Chapter 3, p.37.
2. The goods are usually not measured or weighed. A sack's weight is assessed by the *mandi* merchants by merely lifting it.
3. There is no oil presser in the village. Formerly the farmers themselves made castor oil, although oil pressers visited the village to exchange oil for oil-seeds. Now-a-days, the villagers buy oil in the market than spending time in making it themselves, or to wait for the oil presser whose visit is not certain.
4. The farmers refer to this payment as being equivalent to the annual payment in kind (*adade*) made to the artisan and service specialists. See Karanth (1987a) for a discussion of *jajmani* relations under the impact of sericulture.
5. During 1988-89, a cooperative milk dairy was started which had a daily turnover of about 10 litres of milk during the first few months, and by 1993 it had gone up to about 30 litres.
6. Under the Mysore Silk Worm Seed Control Act, 1951, farmers engaged in sericulture have to be registered farmers who alone can obtain eggs of worms for rearing.
7. See below for a discussion of part-time and; full time unregistered sericulturists.
8. During my recent visit to Rajapura (December 1993) the wages were about Rs. 25 per day without food for a male worker, and Rs. 7 to 10 for female workers.
9. For a very useful description of sericultural technology from a sociological perspective, see Charsley (1982), particularly Chapters 3 and 4.
10. Some of the Talukas known for sericulture in the past are Ramanagaram, Chikkaballapur, Channapatna, Chamarajanagar, Sidlaghatta, Madhugiri, Kollegal, etc. See Charsley 1986; and Ramarao 1954 for a description of sericulture in the past. For more recent developments, see Aziz and Hanumappa 1985; Hanumappa 1993.
11. See also, Buchanan (1807), for a description of sericulture in this region and marketing behaviour in the past.
12. The mediators, referred to as the *palu panchayatdars*, consist of close relatives of the brothers, and a few elders of the village, as also leaders of the political factions to which the brothers belong.
13. The reasons for not taking up sericulture in a given year are many: for example, due to ill health; having some other important work that year like constructing a house, or prepare for a wedding in the family, etc. Often they are also advised by the informal experts to skip a crop or two to overcome the 'evil eye' or a bad phase of luck in sericulture.
14. Doddi, according to the revenue records, is only a hamlet of Kuppe village. But today it appears as though the reverse is true.
15. Villagers consider one as a good farmer (a *kula*) who grows all his ragi requirement in his own fields. However, if a person has no land of his own, the next best is to grow it on land taken on lease or sharecropping basis. In this sense, land is not only "to rule" as Neale (1979) says, but also to be self-sufficient, to achieve the identity of a peasant among the peasantry.
16. If an existing garden is being given for sharecropping, the sharecropper has to provide labour for pruning the plants before taking over the garden for cultivation.
17. See Chapter 5, for a discussion of exchange labour.

5

Landowner and Labourer Relations

In the previous chapter, I have presented an account of how certain aspects of agrarian relations underwent changes while some retained the characteristics of the earlier ones under the impact of sericulture. In this chapter, I shall examine the nature of landowner-agricultural labour relations under the impact of both sericulture and land reforms. There were also certain measures which aimed directly at altering the nature of these relationships, such as abolition of bonded labour, relief measures to the rural indebted etc. These measures ought to have brought about the disappearance of one system of agricultural labour arrangement, and together with other changes in the economy, minimised the interdependence of landowners and labourers. Instead, as shall be demonstrated, here also there is evidence of persistence of interdependence as well as continuity in the systems of agricultural labour as in the past. Such a continuity was not without changes, as shall be seen in this chapter.

Agricultural Labourers: Problems of Identification

There is a difficulty in identifying agricultural labour households in Rajapura. It is conventional to identify agricultural labour households as those deriving all or a major portion of their livelihood from agricultural wages. Conceptualizing would miss out an important section of the rural labour force, namely the "also labouring" *individuals* who form a major source of agricultural labour in Rajapura. Agricultural labour *households* in the conventional sense is almost non-existent in the village.[1] There are many landowning households among whom one or two 'individuals' take part in agricultural wage labour not in keeping with the landowning status of their households. Even while continuing to live with their parents, many young men and widows are found to be working for wages to support themselves or to meet their personal needs. Technically they are members of a household, who, otherwise, do not come under the category of "agricultural labour households" as we conventionally understand them. This phenomenon occurs not only among the 'small and marginal' farmer households, but also among the bigger landowners. Nearly 14 per cent of the 'medium sized' farming households (i.e., owning 5 to 10 acres of land) contribute at least one labourer to the agricultural labour force. However, the marginal and small farmers (i.e., owning less than 2.50 and 2.50 to 5.00 acres respectively) contribute a major share: 57 and 41 per cent respectively. (see Table 5.1).

TABLE 5.1: Households with atleast one labourer

	Landless		Marginal farmers		Small farmers		Medium farmers		Total	
Castes*	No.	%**	No.	%	No.	%	No.	%	No.	%
Okkaliga	13	92.9	8	32.1	7	26.9	1	3.6	29	26.6
Kumbara	NA	NA	1	50.0	NA	NA	NA	NA	1	50.0
Lambani	4	100.0	6	85.7	2	100.0	1	100.0	13	92.8
Odda	2	100.0	3	75.0	NA	NA	—	—	5	71.4
Holeya	2	100.0	3	100.0	3	60.0	2	50.0	10	71.4
Madiga	13	100.0	3	100.0	5	100.0	1	100.0	22	100.0
Others+	7	100.0	NA	NA	NA	NA	—	—	7	85.7
Total	41	93.2	24	57.1	17	41.0	5	13.8	87	47.6

NA = Not Applicable

* Lingayats, Marathas and Koramas do not have any labourers in their caste.

** Percentages here, and in each row refers to the respective caste's proportion of each size class.

\+ *Others* include Madivala, Bestha and Achari castes.

Further it was shown in Chapter 4 that a majority of the landless households are engaged in sharecropping and/or tenancy. As such their major income is not derived from agricultural wages. Landowners offered sharecropping/tenancy to the landless mainly in order to ensure a continuous labour supply from them — a 'labour bank'. A substantial part of the labour requirement of the landowners is met directly or indirectly through such arrangements.

Undertaking sharecropping or tenancy by the landless as well as by the marginal and small farmers affects the availability of labourers in at least two ways. First, the absence of pure wage labourers, many of whom now derive a major portion of their income from being sharecroppers or tenants. They work as labourers only occasionally to support their livelihood. Secondly, availability of such labourers tends to assume exclusiveness and preference in favour of such landowners who have let them a portion of land for cultivation.

In addition to the practice of sharecropping and tenancy, there are also other factors affecting the size of pure labour category. The presence of a forest in the neighbourhood provides an additional, to some a major, source of income. Both the landless and small landowners engage in gathering firewood from the forest for sale in the town. This is highly pronounced among the Lambanis. During 1978-81, they earned as much as Rs. 50 a week. Firewood vendors are able to get back to the village by noon, and still be hired for half-day, or engage themselves in their household tasks or in grazing animals. Secondly, short term and medium term loans through cooperative institutions, and much later through commercial banks have enabled some of the erstwhile labourers to purchase sheep, bullocks, etc. Although not all such beneficiaries have in fact utilized the loans for the stated purpose, nor have all retained the cattle with them for long (see Karanth 1988), there has been a partial withdrawal of some such labourers from the labour force. Thirdly, some other labourers now receive a monthly old age pension (which was Rs.40 during 1978-79, and is now Rs.50 a month) from the state government. With the exception of two or three recipients, the rest, numbering about 30, were ineligible to receive it. As a result of an assured income some labourers have partially withdrawn from labour force. Fourth, a few young boys and girls from almost all castes undertake to work in road construction, repair of tank bunds and so on under the rural employment guarantee scheme or the food for work scheme. Some others who work for a merchant or work for the forest department to lay trenches or in the nurseries. Thus, there has been a few alternative opportunities for labourers other than agricultural, which are more attractive in terms of wages and working conditions.

As a result of all these reasons, the village now faces a continuous shortage of labour supply. Basing on the accounts of the people themselves, this is a marked change from the past when labourers had to wait everyday in a temple yard in the hope of getting hired. There has been a reversal of the situation in which now the landowners are found frequenting the Holeya and Madiga quarters every evening to inform in advance the requirement of labourers for the following day. Landowners are now prepared to concede more and more benefits to the labourers in order to lure them into taking up and completing the assignments. This, however, does not mean that agricultural labourers have been able to get higher wages for their work. The conditions under which labourers come to accept the prescribed wages demonstrate continuity in landowner-labour relationships. I shall describe this as I discuss the different types of agricultural labour arrangements. Four major types of agricultural labour arrangements may be identified in the village. They are: (1) Wage labour, which in turn may be divided into two — (a) casual wage labour and (b) contract labour; (2) seasonal unfree labour; (3) bonded labour; and finally, (4) exchange labour. The distinction between them is an operational one, for the same person may be engaged in more than one type over a period of time.

Casual Wage Labour

Casual wage labourers are hired and paid wages on a daily basis. The wages are paid at the end of the work or as it usually happens, well in advance. A person who works regularly for a particular landowner may obtain wages in advance, or borrow money or grains with a promise to work whenever required. In certain operations, however, wages are paid only on completion of the series of tasks lasting for more than two-three days or even more, or when the farmer gets his produce converted into cash in the market. Thus, a labourer working during harvest operations like threshing and winnowing ragi is paid at the end of all these operations and when the farmer takes the produce home from the threshing floor in the fields. It is customary for a labourer to report for work on all the subsequent days until the last day of winnowing, the day of *kadegana*. On this day all labourers are paid their wages and gifts of ragi. Similarly, female labourers working for a sericulturist in picking leaves and harvesting cocoons, have to report for work on all the days as required by the farmer. They are paid wages for the number of days worked only after the cocoons are sold in the market and the farmer gets his money. Labourers usually borrow money or grain against the wages due. Implicit in these two types of work assignments is that although labourers are casual labourers there is a convention that once a labourer reports for work he or she has to continuously work until the said operation is complete. Apparently landowners will have no money or grain to settle the wages at the end of a day's work until they get their produce sold in the market (in the case of silk cocoons) or bring home the produce (in the case of ragi). Deferred payment of wages under this pretext also ensures a continued supply of labourers on all subsequent days, and they are more accountable. It is a common practice to consider the working day as consisting of two half days. A daily wage labourer may be hired for half a day and be paid accordingly. During the other half of such days a labourer may attend to his own farm work or domestic chores, or take animals for grazing. Ideally, the working day begins at 8.00 a.m. with a break between 12.30 and 1.30 p.m., and ends at 6.00 p.m. By 5.30 in the evening labourers begin to be restless and eager to stop working. It is said that in the past (about 30 years ago, although the past of a villager is timeless), labourers used to report for work by sun-rise, and be working long after sun-set. Farmers now note the change with displeasure that the present day labourers are time conscious and not particular in completing the day's work.

A daily wage labourer is given a breakfast if he works the whole day. A woman worker is not served food even if she works for the whole day in the fields. In contrast, if she works in the house of a farmer she may be given food. Farmers have increasingly preferred to pay for the breakfast, which the labourers take in one of the tea-shops or in their house. Just as landowners complain of the workers starting their working day late in the morning, the labourers also complain over the landowners' paying money for breakfast to be had in tea shops. Labourers claim that the landowners themselves, particularly the women in the household, are lazy, and do not get up early enough to cook. Some even allege that breakfast is served in the landowner's house on such days when there may have been left-over food of the previous night. If lunch is served to the labourer his wage is affected. During operations like sowing, ploughing, harvesting, threshing and winnowing, etc., all male labourers are served breakfast and two meals a day. Some farmers preferred to employ labourers and provide them a meal or breakfast. One woman, mother of a wealthy landowner, explained that farmers consider labourers as their children. Making them work on land without giving food would also make the Mother Earth angry at the farmers. However, there were other important considerations behind such a gesture. If breakfast and lunch are served to the farmer he is likely to be punctual for work, minimum time is lost during lunch hour, and the labourers' attendance is ensured for the post-lunch work. Since there is always some distance to be covered between the worker's house and the field where he has to work, considerable time is lost in commuting for lunch. Female labourers are not served food because they generally work only for half a day (except during harvest season); and they need to go home at least once during the day's work in order to attend to small children and prepare food for others.

Contract Labourers

Contract labourers, either in groups or individually, enter into an agreement to complete a specific task for a fixed amount of wages. This system is known locally as *guttige kelasa*. Tasks entrusted on contract basis are usually those requiring more than one labourer and more than one day's work, e.g., pruning mulberry plants, fencing and digging the mulberry garden, sinking open wells or cleaning the old ones, felling trees, etc. Contract labour arrangement is advantageous for the landowner in more than one way, particularly in the present context of shortage of labour. First, the landowner need not take the trouble of having to find workers for the same task everyday. Secondly, he is relieved of the need for a constant supervision over their work according to time and specific requirements. Thereby, the farmers can devote more time for other operations requiring their personal attention or involvement. The landowner may have to pay a little higher than what he would pay as daily wages for the same work. Thus, petty businessmen, small-time civil contractors, political leaders, brokers in cattle trade etc., usually entrust some of the tasks to contract labourers. This arrangement is also advantageous to such farmers who do not have sufficient number of workers or "helping hands" within the household. On their part, labourers also find contract assignments to be advantageous. They do not accept an assignment on a contract if the payment is less than what they would get on daily wage basis. On the contrary, they now tend to complete an assignment ahead of schedule, thereby enjoy a sense of making a profit. Correspondingly, the landowner will have to his advantage the task completed in a much shorter period. Tasks which are not directly connected with actual cultivation of land, harvesting and post-harvest operations, as also such labour that may require family labour but cannot be performed by them, are being increasingly entrusted to contract labourers. However, this increase indirectly affects availability of labourers for agricultural operations. For, the contract labourers are usually those who also work as casual wage labourers, and contract assignments are generally taken up in groups. While contract labour system saves the trouble of finding labourers everyday until a task is completed, it also affects the availability of casual labourers for other landowners. For example, during the agriculturally slack months of April-May, most farmers undertake to repair their houses, sink a well or get bricks made. For these tasks usually farmers engage labourers on a contract basis. But this is also the season when farmers remove alluvium from the tank-bed and transport it to their fields, preparatory for sowing operations in the coming months. As most of the casual labourers are busy with their contract assignments, other farmers find it difficult to find labourers for removing alluvium. Such a shortage occurs even during agriculturally busy seasons also.

Rajapura not only has its own contract labour groups but also receives a few groups from outside, for sinking wells, house construction etc. While the number of farmers engaging immigrant labourers for agricultural purposes was small during my study, this has increased in recent years due to the persistent difficulty in obtaining local labourers in large numbers for time bound operations. Some of the immigrant contract labourers are horticulturists and Tigala by caste. Most of the immigrant labourers are from Hosapete extension in Magadi town. Because their activities are limited to certain seasons, they are able to complete their work soon and proceed to other villages in the region.

In contrast to the contract labourers, casual labourers lack freedom to determine their working hours or working days. Daily wage labourers are required to report for work on time and keep working till the conventional working hours. In this sense, contract labourers are not subjected to the same degree of super-ordinate/subordinate relations with their employers. They are required to satisfy the broader time limit and quality of work as specified at the time of undertaking an assignment. Failure to comply with it, at least with the time limit given, does not alter the wages. Disputes as a result of non-compliance with all the conditions of a contract either on the part of labourers or that of landowners, are not infrequent. Such disputes are more among the local contract labourers than with the immigrant labourers. A dispute spoils reputation of immigrants which may affect their employment market both locally and, in the long run, elsewhere. Disputes are generally brought before the village leaders who, in most cases, settle the matter.

Seasonal Unfree Labour

A farmer usually advances a loan of money or grain to a labourer on the understanding that the latter will work for the former during the agriculturally busy season like harvest season. To the extent that borrowing foodgrain becomes a binding on the labourer to work, it becomes a seasonal unfree labour arrangement. While a casual labourer also borrows money or foodgrain it is different from seasonal unfree labour since the latter involves an *obligation to work off the interest on such a grain loan*, whereas in the former type a labourer works off the loan. Seasonal unfree labour arrangement is known in the region as *thalavasage*. Under this system a labourer works only to pay off the interest on a loan while the principal has to be returned at the end of the crop season. The system operates as follows: A labourer borrows ragi, usually eight seers or multiples of it from a landowner. In return the labourer agrees to work for one full day or two half days during the harvest of ragi crop for every eight seers of ragi borrowed. The foodgrain borrowed may be repaid at the end of cropping season, or may be deducted from wages to be paid. The system seems to have been in practice in the village since a very long time. Records dating back to 1930's and 1940's maintained by a few farmers show listing of labourers who had borrowed grains under *thalavasage*, and the number of man-days worked towards the interest and principal.

A ritual, and socio-economic significance may be identified in the *thalavasage* system. Grain loans are usually taken during October-November when there is a series of festivals. This period is also a slack season for agricultural labourers, although sericulture has changed this now. The financial and grain positions of the labourers is relatively worse than ever. Hence the need for a grain or cash loan, specially since they need to celebrate festivals. The relatively smaller stock of grain and lack of money seem to have an economic significance in *thalavasage*. During the same month, *mahanavami* festival is celebrated, when farmers of almost all castes offer food and drink to their ancestors. Customarily, labourers make it a point to borrow foodgrain from such landowners with whom their ancestors were closely associated. Close association is in terms of either having been bonded labourers of, or as having been working preferentially for, such landowners. Preferential relations between the landowners and labourers, even if the latter were not bonded, did exist in the past as they do now. Borrowing foodgrain from the former master's household to be offered to the ancestors marks the ritual significance of *thalavasage*. If there has been a marked change in the economic status of children of such former bonded labourers, incidence of their borrowing foodgrain for this purpose is not common. Rise in one's economic status obviously over-rides the ritual significance of *thalavasage*. Indeed, in the present day *thalavasage* system, labourers borrow foodgrain because of economic compulsion rather than ritual or customary significance.

Not all lending-borrowing relations result in temporary or enduring unfree labour relationship with the lender. Thus, persons of equal economic and social status may borrow goods and money from each other without any interest in kind or labour. Even if an interest is levied, the terms and conditions are not the same as in *thalavasage*. The principle of reciprocity of relations discussed in the context of sharecropping and tenancy relationships operates in this context also. Even in borrowing relations between persons of unequal social and economic standing it may not be on the lines of *thalavasage*. Thus, a labourer who works regularly for a particular landowner throughout the year on daily wage basis borrows foodgrain or money without losing his freedom. Nevertheless, landowners may have other means of committing the labourers to work regularly for them, on the basis of which the labourer is lent money or foodgrain outside the *thalavasage* system.

Although landowners complain that the labourers in recent times are not prompt in repaying the loan or working off the interest on borrowed foodgrain, they continue to lend year after year. The major reason for this is to ensure as many labourers as possible to work for them for labour intensive operations. Whether or not a labourer reports for work on a specified day under *thalavasage*, a lending landowner is certain to recover the lent foodgrains or cash sooner or later. As in any other dispute, the village leadership is asked to intervene and pressurise the labourer to repay the loan. Repayment becomes a matter of conscience for the labourers because they would have offered the borrowed grain in worship to their ancestors. By not repaying the loan, they may lose their future credit worthiness. It does not mean that they are always prompt in repayment of grain loans or working off the interest on *thalavasage* loans. It was observed that labourers of Rajapura took more grain loans under the system from farmers of the neighbouring village Palya than from local farmers. The farmers in Palya face an acute shortage of labourers throughout the year, and are always willing to lend money or foodgrains to labourers under *thalavasage* or bonded labour system. Labourers find it easier to borrow from the Palya farmers and delay or skip the repayment of loans or working off the interest. The farmers of Palya are less persuasive than local farmers in demanding repayment. When protested, the local leaders do not show a genuine interest in solving the disputes, for they do not encourage Rajapura's labourers finding work elsewhere specially because of labour shortage faced within the village.

I was informed that when the State Government was implementing debt relief measures during 1975-77, there was an initial decline in *thalavasage* system, as was the case in all private lending relations in rural areas. But the system survives even now: almost all households of lambani, odda, holeya and madiga castes borrow foodgrains on *thalavasage* basis in addition to borrowing on other terms and conditions. Indeed, it was found that the system had grown to such an extent that not only local labourers were borrowing from farmers in other villages, but also labourers from other villages were borrowing from Rajapura's farmers. The existence of such inter-village *thalavasage* relations indicates the continuity of mutual dependence of agrarian categories, reinforced by the shortage of agricultural labourers and inadequate alternative credit facilities for the poor. There were two notable instances of group-*thalavasage* in operation during my study of the village, involving Muslims of the neighbouring Kuppe village. Two groups of Muslim households, consisting of fifteen and seven members respectively, had borrowed hundred seers of ragi from two farmers. During the harvest operations they worked off the interest and later repaid the grain loan. One of the farmers served lunch to the Muslim labourers one day even though it was not part of the agreement.[2]

Bonded Labour System

The system is known in this region as *jita* (Srinivas 1976). Under this system the employer advances a loan of grain or money to a labourer (or his parents) on important occasions like marriage, birth or death in the latter's household. There are two forms of *jita*. The first, is bondedness for an annual wage paid in advance (*sambla geyodu*). It is milder in nature since the labourer does not pay any interest on the money taken in advance. The second is bondedness against the interest on a loan taken, (*baddi geyodu*) while the principal borrowed has to be repaid. In other words, the annual wage is adjusted towards the interest on the loan. Thus, even after working for a year or two, a major portion of the loan remains to be repaid. The willingness with which a labourer offers to attach himself in return for a loan, the urgency of need for a loan, relative lack of alternative sources of credit, credit worthiness, the labourer's physical abilities and material resources etc., determine the form of bonded labourer he becomes. Both the forms of bonded labour have coexisted in the village, although in recent years the former is more predominant.

A study of the history of different households of all castes in the village reveals that a great many of them had at least one bonded labourer during the past three generations. Among Okkaligas, this is applicable mainly to the first and second generation immigrants. Contrary to popular notion, becoming bonded to a farmer does not always mean that such labourers belong to landless households or to lower castes. Although there is a greater incidence of such households having one of its members bonded against a loan, there are many instances of upper caste and landholding households also attaching their members as bonded labourer. Most of them had borrowed money in order to invest in buying land and other assets for the household. One Okkaliga landowner who, today, owns more than 20 acres of land had attached himself for a few years as a bonded labourer. He migrated to the village about fifty years ago with the intention of buying land. Having found that he needed some more money than what he had, he decided to enter into service as a bonded labourer for a few years, and lent the saved money to earn interest. After three years, he was able to save enough money to buy some land. There are many instances even now of a member of a household attaching himself as a bonded labourer for a year or two, so that with the advance payment of his wages the household can afford to build a house, buy cattle, or invest in any other productive way. This does not mean that relatively well-to-do households do not enjoy credit worthiness outside of becoming bonded labourers. This generally occurs among an immigrant household, or a bachelor member breaking away from an undivided family, or to persons with no kin support, and with those who do not like to borrow money from institutional or private sources.

Until the early 1970s, the largest number of bonded labourers were from the lambani, odda, holeya and madiga castes, attached to landowners of Okkaliga caste.[3] Later on, there has been a gradual decline in the incidence of bonded labour among lambanis, holeyas and oddas. Households of these castes have been recipients of government land grants from time to time, which they have been able to retain with them than among the madigas. The lambanis have also increasingly engaged themselves in firewood selling, and along with others, have engaged in sharecropping and tenancy. These alternative sources of income have enabled them to be free labourers.

Another reason for the decline of bonded labour system is that there have been some changes in the landholding pattern which have affected the need for attached labour. In the past, farmers required bonded labourers mainly to have an uninterrupted supply of labour for maintenance of sheep and cattle, clearing the land of trees and boulders, as also to look after the day to day affairs of the farmstead. The average size of landholding was much larger in the past than it is now, thereby reducing the need for an attached labourer among all. Moreover, employing a bonded labourer was a status symbol for farmers of all castes, It was one of the several criteria by which a successful farmer was identified as a *kula*.

A bonded labourer in the past was expected to report for work at dawn.[4] His usual tasks were to clean the cattle shed, transfer night-soil into the manure pit, feed animals, and then proceed to fields after breakfast. He was served lunch in the fields or at his master's house. He remained in the fields until late in the evening accomplishing whatever tasks assigned to him for the day. He brought the animals back from the grazing yard or forest, and secured them in the cattle shed. He was served supper around 8.00 p.m., which he might eat there or carry it home to share with other family members. His breakfast consist of one or two ragi balls and chillies, which were usually left-overs from the previous night, or two ragi *rotis*. His lunch and supper were also two ragi balls each with some curry, and occasionally with some rice. He was given a quota of betel leaves, areca-nuts, tobacco, and *bidis* to smoke. As part of the arrangement, he was also given a shirt, a half pant, a towel, and a pair of sandals. He was also paid for a monthly or bi-monthly head-shave. Over and above these, he could borrow money.

Until the legislation abolishing the system of bonded labour, the labourer and employer generally drew up a written contract on a stamp paper specifying the terms and conditions of attachment. This practice is now almost given up as most of the present day arrangements or oral and informal in nature. Perhaps this may be considered as one of the most significant changes following the legislation. Following the legislation in 1975 the immediate reaction was that some labourers fled the village while in a few others the employers themselves forced their labourers out. Labourers fled because of a rumour circulated in the village that it was a crime to be bonded labourer, and that such labourers would be punished by the government. Most of them returned to the village after a while and came to an understanding with the employers that their loans would be settled in due course either by paying cash or working off as casual labourers. Farmers forced their labourers out because of their fear that they too would be punished for attaching labourers. The written agreements were burnt or hidden, and labourers were instructed to admit, in the event of a formal investigation as was feared, that they were only casual labourers working on daily wages. Some landowners went to the extent of obtaining fictitious vouchers for the daily wages paid to such labourers during the past few years. In due course, the initial impact of the legislation began to wane and the system of bonded labour began to re-appear. Gradually, the employers began to pressurise the 'run away' bonded labourers either to repay or work off the loan. Public opinion being against them and having lost their credit-worthiness the runaway labourers were forced to come to terms with their employers. Government departments like the Block Development Office, and the Revenue Department at the taluk level, which were entrusted with the task of identifying and rehabilitating bonded labourers had not achieved much success because identification was not easy. In Rajapura, only four bonded labourers were officially identified for the programme of rehabilitation, while there had been at least 25 bonded labourers during 1976-77 as my inquiries revealed.[5] Many of those who had undertaken to repay the loans by working as casual labourers sooner or later ended up taking additional loans for their day to day requirements of either from their former employers or others. It was only the immigrant labourers who were able to break the contract and run away without returning for service or repaying the loan.

Sericulture was rapidly expanding, and sharecropping as a mode of cultivation was on the ascendancy during 1975-77. As a result the need for labourers, particularly attached or permanent ones, was more than ever before. The decrease in labour supply due to the run-away bonded labourers, and the initial fear of entering into bonded service on the part of labourers prompted landowners to offer attractive benefits to labourers entering bonded service. Notable among these benefits are the partial or complete waiving of earlier loans, fresh loan or interest free loan, higher annual wages, and most attractive of all, a plot of land for cultivation on sharecropping basis or otherwise. Offering a plot of mulberry garden for sericulture on sharecropping basis has been an important incentive in recent times to lure labourers into bonded service. Thus, one or more of the following factors have encouraged labourers to get attached as bonded labourers: public opinion, pressure from local leaders, need for credit, lack of alternative sources of credit, and fresh offer of attractive terms and conditions by landowners. In addition to such softening of the landowner's attitude, there are also other changes in the pattern of agreement. For example, unlike in the past, a labourer is not required to pay for, or work additionally to make up for, the number of days he had not worked in a year. In a few exceptional cases, a bonded labourer is permitted to earn wages by working for others whenever he is free.

During 1978-79, there were 27 bonded labourers in Rajapura, and two more working elsewhere. Almost all of them were in the age group of 14-27 years. Six had a long history of bonded service while their families also had a similar history. The same year, a 12 year-old boy had been put to service for a loan of Rs. 200 on the understanding that he would work at an annual wage of Rs. 50. The boy's father had worked in the same household since his childhood. Most of the bonded servants were of madiga and odda castes of while there were five okkaliga bonded servants also. The way farmers induce labourers to enter bonded service, or the way labourers enter bonded service indicate that the legislation has not been successful in abolishing the system. What has been abolished is the term *jita* from the vocabulary of the people, since the term now is "annual wage labourer". The system is no different from the earlier ones. The major changes are that it is much more milder than before, and some which are advantageous to the labourers. However, if a labourer discontinues to work having borrowed the money, the principal now attracts an interest.

Exchange Labour System

One of the important features of interpersonal relationships in rural areas is that people attach a value to reciprocity in all their relations with one another. Such reciprocity may be seen in almost all spheres of life. Thus, a farmer feels obliged to return a gift at the time of wedding, or return the help he may have received while celebrating a wedding, constructing a house, attending to an ailing person, or in funerals. There is also a tendency to match person to person, form to form, quantity to quantity in such reciprocal relationships. Generally, these reciprocal relations transcend caste and class boundaries, although intra-class reciprocal relations are common. One form of such reciprocal relations is well organized in the sphere of agricultural operations, under the system of exchange (*muyyi*) labour.

Certain agricultural operations require a large number of workers, which have to be completed at a stretch making use of favourable climatic conditions. They are, for example, ploughing, sowing, weeding, transplanting, and harvesting. Sometimes, these operations also require large number of draught animals, ploughs and other implements, which all farmers may not possess in sufficient numbers. Labour supply during these seasons is also inadequate since almost all farmers, including the small and marginal farmers, will be engaged in the same operations in their holdings. In order to overcome the shortage of labour and the need to finish the tasks at a stretch, landowners pool their manpower and other resources to work on each other's land. Exchange labour group is a result of such pooling (see Chauhan 1967; Oommen 1972; Breman 1974; and Epstein 1962).

Over the years the importance of exchange labour groups has increased for reasons discussed above, and since sericultural activities also demand attention of farmers throughout the year. This increased importance goes to indicate the continuity of traditional character of interdependence in agrarian relations. In addition to solving the problems of shortage of labour, it also helps such farmers who cannot afford to pay for the wages of hired labourers. Generally, farmers owning adjacent lands make up such exchange labour groups. Having pooled their labour, they work on each other's land by turns. If one puts in more number of man-days for the others while he himself required lesser number of man-days, the difference is paid for as wages. Such wages are not paid if the group is formed by close relatives or friends.

There are a number of issues over which an exchange labour group experiences disharmony or eventual disintegration. Some of them are, for instance, the question of which member's land has to be worked on a given day; premature withdrawal by some from the group; not settling the wages to be paid for extra work taken in, and so on. Having been in practice for a very long time, certain conventions have been evolved for the smooth running of the exchange labour group. Certain reasons are recognized as legitimate excuses for premature withdrawal from a group, such as death, birth or marriage in the household, or severe illness to the participating members. The conventions are, by and large, adhered to by the members since disqualification in one group may mean reluctance by the other groups to give membership to such a person.

The group has one of its own members as the leader who organizes and runs the group right from its inception to dissolution at the end of a season. The leader, referred to as *mestri*, decides whose land is to be worked on a particular day. He also maintains the log book to record the contribution of labour by each member, and calculates the wages due from each member to the others. As the participating members are also landowners, collection of wages becomes a delicate matter; therefore, the leader functions as an useful mediator for them. As a privilege, the leader need not actually work the way others do. Instead, he supervises the work and makes necessary arrangements like supplying drinking water and refreshments to the members of the group. His land is the first to be worked upon by the group; in the days to come the leader is busy with the day-to-day administration of it.

Two types of exchange labour group may be identified. The first is a pure exchange labour group, which dissolves itself immediately after completion of work for all the participants. Generally, it consists of well-to-do farmers whose family members do not normally work as wage labourers. By and large, the group consists of members owning more or less the same extent of land. The second type of group persists even after the completion of exchange labour among the members, which now begin to work for non-members for wages. Normally this type consists of members who are marginal and small farmers as also the other households in which members work as wage labourers. Irrespective of the demand for their labour, this group completes operations first on the lands belonging to the members. It is not uncommon for a non-member landowner to offer higher wages or other incentives (such as lunch for the workers, or a 'bribe' to the leader) to lure them to work on his land. This is done usually to supplement the work being done by another group of which he or members of his household may be participating.

During harvest of ragi crops there are nearly 20 exchange labour groups of varying numerical strength. These groups are organized generally by young men who can read and write. Some households owning large holdings encourage as many of its members as possible to take part in as many groups as possible. The advantage being they can get maximum number of workers on the same or successive days. Quite a few landowners employing bonded labourers sent them to participate in these groups. Likewise, those who are unable to take part in the working of a group for a day or two may hire a labourer to represent them in it. A mutual exchange of labour among households is not uncommon wherein a household may send an additional member on behalf of another, in reciprocation of an earlier gesture of this kind.

There are exchange labour groups which are exclusively by women. These are formed for such operations requiring female labourers, as in transplanting and weeding. However, such groups are led by males for they are considered to be capable of better management and accounting. Exchange labour groups for harvest operations are by both men and women, although men predominate them. Numerical predominance is seen particularly in the pure exchange labour groups, which may be indicative of people's status consciousness and a behaviour specific to a given economic category as expressed through prevention of women's participation. It may be pointed out here that women do take part as workers (hired or unpaid family labourers) in all agricultural operations, and in this case, on such days when exchange labour groups work on the household's land.

The size of exchange labour group is much smaller and more selective for ploughing and sowing, and for post-harvest operations. In these operations, two or three farmers come together by pooling not only their labour but also other resources such as cattle, ploughs, tools and implements, and in the case of post harvest-operations, the space for threshing floor. Since such groups are generally composed of a household with intensive and intimate relations, "wages" for differential labour or other inputs are rarely paid.

What is the significance of persistence of mutual exchange labour? The description above points out that, contrary to generally held belief (see, for e.g., Frankel 1971; Alexander 1975), interdependence still persists among farmers and different sections of the village. This becomes even more important when it is pointed out that exchange labour groups consist of members of diverse economic and caste status. The participating members are often not even on speaking terms with each other following disputes over different issues and membership in politically opposing factions. Notwithstanding such disputes among themselves, the members work on each other's land as members of mutual exchange labour group. Moreover, the spirit of interdependence and reciprocity of relations is encouraged by the need for labour. This is also because every farmer feels the need to be accepted in one group or the other rather than being isolated.

Wage Free Labour

In addition to the above mentioned four types of agricultural labour arrangements, there is yet another, i.e., wage free labour. This type has not been included in the four-fold typology above as it is not a systematically defined one and usually involves only a fraction of the working day. Such wage free labour is rendered out of some kind of obligation to landowners by labourers who are situationally vulnerable, or even by persons of lower economic and social status. It need not be confused with *begar* system of labour where there is an element of customary relationship involved (see, Thorner and Thorner 1962). Thus an okkaliga housewife of a well-to-do landowner can get a wooden log cut into pieces by a madiga landless labourer who may be passing by the house. Rendering such services is not uncommon for the former bonded labourers to their ex-masters. On my way back from Magadi town one evening, an accompanying ex-bonded labourer of the headman's household, who had been sent on an errand to Magadi, narrated how members of the dominant caste extracted work without paying any wage. He seemed to have been amused by the fact that such obligations are expected even by the landless simply because they are members of the dominant caste.

He went on to elaborate the vulnerability of his caste fellows. It appears his father had borrowed Rs. 100 some years ago from a landowner, for which my informant had been a bonded labourer for a few years. It had been a bondedness to work off the interest on the loan, and it had taken five years to free himself. Now he had been a sericultural sharecropper with the same landowner, and for the past six months the latter owed him about Rs. 600. This amount was due to him towards his share in one of the crops of cocoons sold in the market. He referred to the paradox in the situation, in which for Rs. 100 borrowed by his father he was a bonded labourer for five years, whereas six times that money is now due to him for which no interest is being paid. That morning he had gone to the landowner's house to request for payment of the money due to him, when his wife had assigned him to fetch some things from the town.

Wages

Vulnerability of agricultural labourers may be seen further in the wage structure. The periodic fixing of minimum wages by the government has, by and large, remained on paper. A report in 1980 had showed that the state governments were not effectively imple-menting the Minimum Wages Act, 1948. With reference to Karnataka the report noted that the minimum wage fixed for the state was between Rs. 3.25 and Rs. 5.50 per day with meals, or Rs. 7.00 without meals, which were far below the bare minimum consumption needs of agricultural labour households (cf. *EPW*, 1980, Pp. 1044-45).

In Rajapura, none was aware of the Minimum Wages fixed by the state. Although it was envisaged by the Prime Minister's 20-Point Programme that equal wages be paid to male and female labourers, discrimination between them was found to exist in wages and working conditions (see Table 5.2). Wages are generally linked to whether or not a labourer is served meals by the landowners. As a rule, female agricultural labourers are not given a meal or breakfast. It was explained to me that by not serving food to a female labourer an opportunity was given to her to go home and to attend to some chores like cooking food and paying attention to children. One person with a liberal outlook pointed out that by allowing women labourers to go home for lunch, the landowners give her a break from work. However, they are served food during the post harvest operations like winnowing and cleaning the grain. There are some landowners who do serve lunch to their female labourers so as to save time and ensure the latter's attendance for the second half of the day. Such an act is purely voluntary and does not affect the wage.

While there is a coexistence of wage payment in cash and kind, there is an association between cash or kind wage payment on the one hand, and the nature of operations and the sex of labourers on the other. The mode of payment in almost all operations to male labourers is in cash, except when they demand it in kind. There are some operations like weeding for which payment is made in kind to both male and female labourers. The wage payment is in kind and cash for transplanting paddy saplings and harvesting paddy. Labourers receive

TABLE 5.2: Wages for Different Agricultural Operations (1978-80)

Operations	Wages for: Male labourers With meal	Male labourers Without meal	Female and child labourers Without meals
1. Ploughing dry fields (cash)	3.00	—	NA
2. Weeding (kind)	NA	NA	2.00
3. Sowing (cash)	2.00	3.00	NA
4. *Kunte** (cash)	2.00	3.00	NA
5. Harvesting and post-harvest work (cash)	3.00	—	3.00@
6. Polughing wet fields (cash)	4.00	—	NA
7. Transplanting paddy (cash and kind)	NA	NA	3.00
8. Transplanting ragi plants	2.00	3.00	2.00
9. Harvesting paddy (cash and kind)	4.00	—	3.00
10. Digging mulberry garden or groundnuts	2.00	3.00	NA
11. Picking mulberry leaves	NA	NA	1.00
12. Picking avarekai	—	1.00	1.00
13. Scrapping and collecting alluvi	3.00	4.00	
14. Riding bullock cart**	10.00	10.00	—

* *Kunte* is an operation of weeding, done with the help of bullock drawn iron blades that uproots weeds when the food crop is still growing (after two to three weeks of sowing). This is done in fields where ragi is sown with the help of *kurige*, (that is, which is sown in the lines and not haphazardly). Both manual and bullock drawn weeding is performed depending upon the rains and the growth of weeds, as also the type of sowing the seeds.

@ During harvesting and post-harvesting seasons the labourers are provided with meal and coffe/tea for both male and female labourers

** If the farmer rents in a bullock cart and if the owner of the cart himself rides it with the use of his animals. However, if a laboûrer is employed to ride the cart, he is paid as in. No.13.

payment in cash only for work done during the harvest of ragi, whereas they are paid in kind for all post harvest operations like cleaning and winnowing. During the early part of the agricultural year and during the post harvest seasons the grain stock is comfortable, and therefore wages are paid in kind. This pattern of wage payment has certain economic implications also. When the landowner pays the wages in the form of grain, he will be spending less money in relative terms. This is because the price is low due to influx of foodgrain in the market. When wages are paid in cash, the landowner is again saving money in relative terms. The buying capacity of currency has come down as a result of higher price due to decreased supply of foodgrain.

From the labourers' point of view, this has certain advantages. At the beginning of agricultural year he does not have a large stock of foodgrain. He accepts, therefore, readily the wages in kind. By doing so he will gain in relative terms since now the price of grain is high in the market. During the latter months, and during the post-harvest seasons, he has a good stock of grain. He is, therefore, tempted to accept wages in cash which he can spend on paying wages to the labourers whom he may hire, for visiting places after harvest, attending and celebrating weddings etc. Another important source of grain for him during harvest operations is noteworthy, since it indicates to one of the survival strategies of the poor. Labourers pick up the fallen grain-stalks (*theney*) from the fields which may fetch him 3 to 4 seers of ragi per day. Most landowners do not object to the labourers picking up fallen grain stalks, specially if it was done while they were working for wages in their land.[6] On an average a household with three labourers and two dependents can save about 60 to 70 seers of ragi for the agricultural slack season.

Female labourers are paid wages in cash for only harvest operations and all sericultural labour. In the latter, payment is made only after the farmer receives his money from the cocoon marketing officer or private buyers. Labourers, male and female, generally borrow money in advance. Women borrow more often to spend on household expenses, since generally male members spend their earnings for their own needs. It is quite common in the holeya and madiga quarters to witness in the evenings a father and son, or a husband and wife, quarrelling over money earned and spent unaccountedly.[7] Most labourers maintained a credit account with the local merchants. These merchants are seen in Holeya and Madiga quarters during early mornings or late evenings pressing for repayment of a long due grocery loan or settlement of a bidi-tobacco account.

It is also common for male labourers to borrow money from landowners with a promise to work on any specified day. The landowner reminds the labourer a day or two in advance, and sometimes goes to his house to page him for work. Such incidents of the landowner visiting labourer's house, or the tea-shop (*hotelu*) where the labourer would be gossiping, are becoming a common sight in recent years. Generally, labourers keep up their promise to work though not on the day as specified at the time of borrowing money. Thus the landowner's investment does not generally go waste though it may involve some friction and hot exchange of words.

Finally, I shall refer to the lack of bargaining power of labourers to demand higher wages. In this respect landowners act as a more united front than the labourers. I have already referred to the short supply of labourers in general and during certain seasons in particular. Yet, there seems to be very title bargaining by the casual wage labourers for higher wages. There is some bargaining involved in contract wage labour, and in fixing the annual wage under the bonded labour system. Even here, the landowner has a better say in the matter than the labourer.

Landowners compel the labourers to accept the existing wage rates over which the latter express their dissatisfaction. During 1978-79, teams of harvest labourers claimed higher wages than the existing one of Rs. 3.00 per day. Their basis for such a claim was that the immigrant labourers were paid at the rate of Rs. 4.00 per day for the same work. During that season one landowner paid higher wages to get his work done faster than others. The local labourers began to press for the same wage rate from the rest of the landowners in the village. The matter was referred to the headman and the former *Patel*. After an emotional appeal and heated arguments on the part of landowners and labourers (many of the latter themselves being landowners), the two leaders declined to let the wages to be increased. They pointed out to the labourers that the "immigrant labourers did not have any obligation towards the village, and *vice-versa*. Landowners were only supplementing their labour requirement by hiring the immigrant labourers; otherwise, they may suffer a considerable loss. As the immigrant labourers had come from Magadi, the higher wage would compensate for their travel expenses". The local labourers were given the option of either accepting the existing rates or face a boycott by the landowners for all operations in future.

I am inclined to believe some of the explanations given by the labourers in meekly accepting this verdict, some of which have been found to be true elsewhere. A few more need to be highlighted here. First, the labourer's perception: lack of unity among them, vested interests among the landowners, and a false sense of obligation to the village. Among the other factors, I think the role of the local leaders is important in not only the affairs within the village but outside it and in the bureaucracy. The headman was and has been since 1970s an important leader at the taluk level. To have him against would also mean loss of his patronage in all other respects too. There may have been a police case, an application with the credit cooperative society for a loan, an application for a grant of house site, loan to build the house and so on. All these require the good offices of the local political bosses.

Concluding Remarks

My description of agricultural labourer and landowner relations covered the main types of labour arrangements, and a brief account of agricultural wages. There have been state initiated measures which may lead one to expect more impersonal and monetised relations between the landowners and labourers, as well as a declining interdependence between the two. The measures towards a minimum wages to be paid, equal wages to men and women, etc., did not have any significant impact. The village leaders used their positions to safeguard the interests of landowners rather than that of landless labourers who are indeed a minority in the village. The increased demand for agricultural labourers has not increased their bargaining power. Continuity of exchange labour system is one of the main factors which comes in the way of better bargaining ability of labourers.

Some of the institutions that prevailed in the past such as bonded labour system, *thalavasage* system continue to flourish even after changes in the economy and introduction of welfare measures. The institution of *thalavasage* was affected only for a few months following the introduction of debt relief measures. But soon labourers continued to approach landowners for short-term credit requirements. Bonded labour system was also in existence with a sizable number of labourers attached for a year or more, although it had undergone some changes like more annual wages, no interest to be paid, a plot of land for cultivation or an opportunity to the labourer's household members to sharecrop in sericulture.

Interdependence between landowners and labourers had not completely disappeared. The fact that some labourers get credit is a sign of 'help' from the landowners. This is particularly important when one notes that none of the 'identified' landless labourers or agricultural labourers had been benefited by way of securing institutional credit or the fact that not all the former bonded labourers were identified for rehabilitation. The four identified bonded labourers were able to get help only in 1982, nearly six years after their supposed release from bondedness, by which time they had already re-entered the service of their former masters. Many of the former bonded labourers continue to maintain their good relations with their former masters, typical of the patron-client relationship in rural India. Many labourers were known for their working preferentially for certain landowners even if it was for daily wages. These suggest enduring aspects of relationships between the two agrarian categories.

The coming of a cash crop, sericulture, did not alter the wage structure either in terms of mode or quantity. Sericulture drew many women labourers into the labour force. Sericulture did not any significantly alter the interpersonal relationships between landowners and labourers. Just as agricultural labour, female and male labourers engaged in sericultural activities also wait till the employers sell the produce to get their money. Delayed wage payment also ensures that the labourers continue to work for the same landowners. This is further enhanced by the frequent borrowing of money as well balance of money with the landowner.

Notes

1. Indeed, the 1971 Census data for the village showed only 2 agricultural labourers.
2. See Bardhan and Rudra (1980) for an account of an arrangement similar to *thalavasage* in West Bengal.
3. Generally, an Okkaliga bonded labourer served only in the household of his own caste, or castes above his in the hierarchy.
4. Although some labourers used to stay in the master's household, many young labourers were exempted from this requirement. Since almost all young bonded labourers would have entered the service following the borrowing of money or grain for their marriage, and therefore were newly weds, the landowners usually did not insist on the labourer living in the same house with them. Notions of pollution was also an important consideration in relaxing the rule of residence.
5. The four identified labourers had to wait till 1982 when they were given sheep worth Rs.1,200 as part of the rehabilitation programme.
6. Some labourers even spend their evenings digging rat-holes in the fields to collect the grain stalks deposited there by the rodents. Labourers and many youths from landowning households trap field rats under the hay-stack during nights. The lot is shared by the group, and cooked at home. Summer months are also the season for many young men to venture into the forest at night to shoot rabbits, wild pigs, chicken and so on.
7. An Odda household sells liquor, unauthorizedly, in the village. During 1978-81 this liquor shop was patronised mainly by the former untouchable castes, although a few Okkaligas were discrete in their custom to the shop. Even the Holeyas and Madigas were secretive about their drinking habits, at least within the village. During a recent visit (December 1993), I was surprised to see the openness with which the liquor shop was being patronised by not only the earlier customers, but also by the members of the dominant caste and others. Elders blamed the leaders and their style of electioneering in the past few elections which tacitly gave license to the drinking habits of the people.

6

Conclusions

It is generally believed that the policies aimed at macro changes fail to have a uniform effect over the country as a whole, because of the variations in agrarian relations and other social, economic, political and historical factors which are specific to the different regions. This study examined the impact of some state initiated measures on micro level agrarian relations, and to some extent at the regional level. In this concluding chapter, I address myself to two questions: what has been the overall impact of these state initiated programmes upon the traditional agrarian relations? What bearing does this process has on our understanding of the process and nature of social change in rural India? In order to answer the first question let me summarise the discussion so far made in the preceding chapters.

With regard to land reforms, it was observed that by the time reform measures were announced and implemented many of the absentee landowners, with their awareness of the impending reforms, had sold their land to whoever came forward with competitive offers. Those who bought such lands were, by and large, the immigrants who had come to settle in the village in the hope of buying land. These immigrants had made use of their caste and kinship network for not only migrating into the village, but also in finding land for tenancy cultivation, mortgage, and finally buying land. The dominant okkaliga caste had a major share in this process. Thus, there was one important change in agrarian structure as a result of land reforms: transfer of land from absentee landowners (mainly from the non-okkaliga castes) to the resident Okkaligas. At the same time land reforms can also be seen as having reinforced the caste and kinship ties which were made use of for the purchase of land. In keeping with the general trend elsewhere, there had been large scale eviction of tenants from land in the wake of land reforms. In the process the hitherto existing power structure was reinforced as mostly leader's help was taken in eviction of tenants and in disposing off the land through sales outside the framework of land reforms. By and large, land was transferred to Okkaligas, and/or supporters of the leading faction to enhance the power base of the hitherto powerful and wealthy landowners. Only such cases of tenancy that could not be disposed off in the above manner found their way to the land tribunal. The scope of power and influence of local leaders included the bureaucracy too. For the leaders this was one more opportunity to demonstrate their power and influence with bureaucracy. The officials were not free from prejudices, and in return for favours of transfers, bribe, etc., sided with the dominant caste leaders. Non-official members of the land tribunal were those nominated by the state government, recommended by the Members of Legislative Assembly, or other leading politicians at the taluk level belonging to the ruling party. In Rajapura, the dominant leaders were staunch supporters of a former M.L.A, whose nominees were in the Tribunal. The headman from the village served as its member for more than ten years.[1] Indeed, the local leaders were often approached by neighbouring villagers to speak in their favour to the members of the tribunal. The members, in turn, were not free from considerations of party and factional loyalties. In Rajapura there had not been any need for such an intervention by the tribunal members. Land records had been suitably altered with the connivance of the leaders and officials. If a tenant was to be favoured *pahani* records were altered to accommodate him; and if a landowner had to be favoured, the entries of tenant's names were removed. A bribe of Rs. 100 to Rs. 200 to the village accountant took care of this.

More than the genuine cases of tenancies, applications to ratify illegal land transactions were in a majority before the land tribunals. There had been a substantial number of multiple applications claiming confirmation of occupancy rights over the same plot of land.[2] Most landowners had induced others to make multiple claims so as to complicate and delay the process of settlement, by which time a genuine tenant may give up hopes of settlement or be forced to arrive at an 'enforced compromise'. It was observed that counter applications for the same land were generally by resident tenants for resident landowners' land. A majority of the cases that were 'pending' before the tribunal for settlement during 1978-81 were of this type, understandably because of the complications.

More than three-fourths of the applications before the tribunal in the taluk were for lands of absentee landowners. They were, by and large, from non-peasant castes, such as Brahmans, Marathas,

Vaishyas, etc., living in urban areas in and outside the state. Further, a majority of the applications were meant to ratify a land transaction made earlier. Often, this was done in order to save the fee on registration, or if such a transaction violated the existing land transfer legislations. Given this, there is a need to look at official statistics regarding confirmation of occupancy rights to the tenants with some caution.

Two important forces were at play with regard to eviction of tenants of resident landowners. First, the tenants lacked an awareness of procedures and legal provisions. They were invariably dependent upon the local elites and professional "application writers". Local leaders, therefore, determined what the course of action on the part of tenant should be. Leaders have a variety of means to bring pressure upon tenants, failing which they try to complicate the matters through counter-applications, influencing the tribunal members, pressurising officials to adopt delaying tactics, influencing the witnesses and finally manipulating the records. The Karnataka Land Reforms Act had provision for the members of the tribunal to go on a spot inspection, and such visits were not free from influences of the vested interests.

Secondly, the network of relations between the bureaucracy and local elites played a crucial role in the implementation of land reforms. These leaders are treated with special reverence when they visit the government departments. In turn, the leaders kept the officials in good humour by supplying them with firewood, foodgrains, by routing the bribes to them, and in helping them to get them transferred to a place of their choice. Officials who tried to remain neutral, or independent of the leaders were "harassed" through transfers, petitions against their style of working, and even getting them "caught red handed while accepting a bribe". Thus, ranging from the village accountant to the Tahsildar (taluk Magistrate), all the bureaucrats were under the influence of the leaders. On their part, thus, the tenants needed the good offices of the bureaucrats from the village accountant upwards. Being obligated to the leaders the officials were willing to favour the leaders than the tenants. While emphasizing the role of leaders I do not imply that there are no factionalism among leaders. But, as was observed, officials generally tend to side with the leaders of the ruling party. The net result of these two factors, namely, lack of awareness among tenants and the role of political leaders was that many tenants of local landowners were unable to seek legal protection under the law.

Another aspect concerning tenancy with local landowners needs further evidence from other areas. In Rajapura (and also in the surrounding villages) it was found that this type of tenancy in the past had not been so much on a 'commercial', 'formal' or 'capitalist' pattern. In other words, the act of leasing out land by a resident landowner was not always and necessarily for profit. It was found that many such tenants had been cultivating a small piece of land to grow a particular crop for a year or two. These arrangements were, by and large, oral and informal. While initially, as a result of land reforms, there had been a decline in this form of tenancy, it was found to be revived by 1978. This revival also demonstrates the continuity of a form of 'traditional' mutual dependency relationship. During 1930-50 there had been caste and kinship sponsored immigration and land transfers, likewise in recent years the resident landowners had been leasing out land to their relatives who held insufficient land to meet all or part of their foodgrain requirements. Further, landowners now resort to oral and informal leasing out to the labourers so as to ensure a continuous labour supply from such labour households.

Thus, it was found that land reform legislations had failed to effect the stated goals of tenancy reforms, security to tenants, and later, conferring occupancy rights and ownership rights to the tenants. On the contrary, land had been sold to non-tenants and had caused substantial eviction of tenants. For the evicted tenants there was a hope to buy land which was coming into the market from absentee landowners. Rather than incurring expenses on legal formalities they chose to invest in buying such land if they could.

The coming of sericulture was the second factor accounting for some changes in agrarian social structure. With the gains from sericulture some farmers had bought land, constructed new houses or renovated the old ones, sunk wells, and in a few cases, educated their children in colleges and universities. Some of the notable changes were: in land use pattern, increased irrigation and demand for irrigated land, creation of a new class of workers and roles specific to sericulture, increased demand for the services of artisans, increased participation of women and child labour, and an increased interaction with neighbouring villages. However, the new cash crop had certain constraints in terms of shortage of labour, risks involved, fluctuations in the prices of cocoons, and inadequate water resources — all of which came in the wav of undertaking large scale sericulture. However, the farmers did attempt to intensify sericulture by adhering to certain modes of traditional agrarian relations. Important in this respect was letting out land on tenancy or sharecropping in sericulture. While sharecropping and tenancy enabled the landowners to earn more from sericulture, it also enabled them to ensure an assured labour supply for both sericulture and agriculture. Further, sharecropping and tenancy in food crops was found to be reinforcing caste and kinship ties which had been witnessed prior to the coming of land reforms. Thus, contrary to the generally held view, land reforms and a new commercial crop had reinforced traditional relations between different sections of agrarian social structure.

Land reforms and sericulture had contributed also to the continuity of some other aspects of traditional agrarian relations. Although not discussed separately in my account so far, I shall briefly refer to two of them, namely *jajmani* relations and informal credit relations.[3] The new landowners and immigrants had needed the services of different types of artisans and specialists. By owning land they came to possess the necessary qualifications to obtain such services under the *jajmani* system. Sericultural technology is such that it requires various appliances that are made by local artisans like the basketmaker, carpenter, blacksmith and potter. Thus, sericulture too strengthened the need for the services of artisans, and within the framework of *jajmani* relations. In this context too, contrary to the generally held view, a commercial crop did not reduce the importance of a traditional network of relations involving customary exchange of goods and services.

When I use the term 'traditional' in the context of *jajmani* relations I refer specifically to Rajapura's tradition. *Jajmani* system in the village even in the past had departures from the all India pattern which one may discern from the works of other scholars. In Rajapura the practice of hereditary succession to the office and craft exclusiveness were not strictly followed. Just as local farmers had sponsored immigrants in the past, a few patrons of the washerman and black, smith had also sponsored new ones and inducted them into the network of *jajmani* relations. In other words, traditionally also *jajmani* relations had undergone changes and the system had been adaptable to such changes. As a result of sericulture and land reforms certain changes were observed. There had been a decreasing importance of the customary formula in determining wage payment to the specialists[4]; coexistence of cash and kind wage payment; frequent change of patrons among the competing specialists; and relatively increased discontentment as a result of non- or irregular payment of wages. The last two changes may appear to be signs of breakdown of *jajmani* relations. But, as was observed during 1978-81 and in subsequent years, local leaders had intervened in arranging a compromise so as to encourage the patron farmers to make payments on time. It was also found that members of respective castes had been willing to take up craft/service if an incumbent gave up the office. In fact, as in the case of musicians and one washerman, they had extended their services to new villages under the *jajmani* system in addition to the other villages they were hitherto serving.

Further, continuity of *jajmani* relations is also to be seen in terms of the conditions of serving each other, for instance, continued restrictions on serving the former untouchable castes, even on cash payment. Primacy attached to the notion of village as a corporate entity is to be noted here, for such services were made available to them by the same specialists against cash payment in places like Magadi and elsewhere they may have a shop. A change in the economic status of the former untouchables like Holeyas and Madigas has also failed to bring about a change in this respect. Nor has there been any significant attempt so far on the part of such castes to protest and demand services from such specialists in the village. Educated and urban employed youth play a passive role in this regard, who try to avoid any confrontation with the other castes. Such efforts, they fear, would only make things worse for their parents and kin living in the village (see Karanth 1981).

Continuity of 'traditional' interpersonal dependence, if not *jajmani* relations, is also to be seen in the case of extending notions of it to the gifting of hay to the visiting merchant and to the forest official. The villagers' relations with the visiting merchant is expected to be market oriented, commercial and impersonal. Instead, it has come to acquire personal and one of mutual interdependence in which both borrow from each other at times. Likewise, with respect to the forest guard the relations are expected to be impersonal and bureaucratic. Yet they are found to be intimate, personal, and in some ways interdependent. In this context I may also cite an example of the continuity of relations with the former absentee landowners. There are more than a dozen farmers who had years ago bought land from absentee landowners living in places like Bangalore, Kunigal, etc. Even now such formers take with them a quantity of ragi, fruits, ghee, and vegetables as gifts to such former landowners. This is often in spite of misunderstandings at the time of land transactions, as for instance, improper land measurement, or increase of the land price than what was quoted earlier, etc. In return for such gifts they receive gifts of steel vessels, woollen blankets, clothes and occasionally credit. The sense of loyalty and obligation on the part of the farmers is in return for the landowners selling the land to them. On their part the landowners also appreciate to receive some gifts of grain and crops grown on land owned by farmers.

It is appropriate here to mention one more evidence of 'continuity' in traditional aspects of agrarian relations which I have dealt with separately. This refers to the mutual trust and confidence in intra- and inter-village cash transactions. During the annual cattle fairs held in different places to which farmers from this village go, farmers sell their cattle to total strangers from elsewhere. Such transactions take place between farmers belonging to villages of different taluks. The money is often collected weeks later either at a cattle fair elsewhere, or through a messenger. Only the cost of the rope fastened to the cattle is paid/received at the time of transaction. I must, however, qualify this observation that mutual trust of this kind have been misused *both in the past and present.*

Continuity is also to be seen in the context of the postponement of wages to the labourers in sericulture, during the harvest operations and in the payment for goods sold to the visiting merchants, agents of licensed seed preparers, wholesale merchants in Ramanagar vegetable market etc. All such delayed payments suggest not only mutual trust but also endurance of relations. Thus, there are some evidences suggesting continuity of 'traditional' aspects of agrarian relations even after the introduction of a cash crop, which, however, has not been without changes.

With regard to credit relations it was seen that very often the motive for moneylending was to ensure a labour supply than merely as usury. The terms and conditions in money/grain lending varied according to the socio-economic status of the debtor, which was in keeping with the principle of reciprocity. The institutional measures to provide credit had suffered a set back as a result of faulty identification of the targeted beneficiaries, which was mainly due to the nature of land records upon which identification was made. The Primary Agricultural Credit Cooperative Society (PACS) and the Primary Land Development Bank (PLDB, formerly the Land Mortgage Cooperative Society) had been catering mainly to the big landowners, with a very high incidence of defaulters and passive participation. The PACS had been reorganized in such a way as to cater to the credit needs of nearly 32 villages, more than half of which had no transactions with it due to the distance involved, or being unaware of the reorganization of the PACS. Both the unpopularity and misuse of institutional credit was found to be due to the misuse of positions enjoyed by the elected representatives to these institutions, officials, ordeal of obtaining credit in times of need, and the social 'humiliation' faced at the time of recovery of loans.

As an important change in credit relations a reversal in the direction of flow of credit was found in recent times. Though this was not on a large scale, it indicates the changing positions of the formerly poorer sections of rural society, which in any case shows the continued vulnerability of such people. Small and marginal farmers, and members of lower castes like holeyas and madigas were found to have lent money to the bigger landowners and to the members of dominant castes. However, neither the Holeyas nor Madigas have been able to enhance their social status as a result of changed economic status. Segregation and discrimination continue to be practised in affairs concerning these two castes, as for example, denial to draw water from the village wells, denial of entry into temples and houses of upper caste members, and denial of services of washerman and barber, etc.

My data on incidence of indebtedness show the predominance of private sources of credit, while many bigger landowners have borrowed from both the private and institutional sources. The impact of debt relief measures had not been felt (none in the village having formally sought any relief), although initially there had been a decline in loan transactions with private sources. Given the inadequate alternative sources of credit private moneylenders soon began to be approached for fresh loans. As a result of all these factors, viz., faulty identification of the beneficiaries, corruption and delay in sanctioning loans, social humiliation to be faced at the time of recovery of loans by the officials, and lack of adequate alternative sources of credit, private sources for credit continues to be a significant one. This, in turn, leads to the continuity of the dependence of agricultural labourers and other borrowers upon the rural moneylending landowners. Moreover, there has now been an increasing rethinking among scholars on the role of the rural moneylender which needs to be taken into account while considering the policies on rural credit (see, e.g., Michie 1973; Hill 1982, and Karanth 1988).

Finally, let me turn to landowner-agricultural labourer relations.

Welfare and rural development programmes like providing food for work, employment guarantee programme, old age pensions, etc., had also resulted in a shortage of labour supply in agriculture, while sericulture had intensified labour use. Yet wages had remained relatively unaffected, nor had they any impact upon the bargaining ability of the labourers. When compared to an ideal of glorious and harmonious past as often narrated by the villagers with regard to landowner-labourer relations, one could see considerable changes in terms of strain and stress in relations between the two. Frequent denial of credit and other such help by the landowners, a near organized attempt to keep the wages low, increasing dependence on migrant labour etc., are some of the changes on the part of landowners. On the part of labourers there were instances of not repaying in time a loan, or working off the loans, reduced working hours, 'lack of commitment', were some features which may be considered as changes. There is no evidence to suggest that these are peculiar to the contemporary Rajapura: incidence of this nature was as much a part of the agrarian social structure in the past as they are now.

In terms of labour arrangements, contract labour was found to be becoming more popular because of its advantages to both land-owners and labourers. In the past, this form of labour was restricted mainly to the tasks (such as fencing, felling a tree, sinking a well, etc.) which were not directly connected with cultivation. It is now practised even in main agricultural tasks such as weeding, transplanting, or harvesting, and in sericulture. Two of the most important aspects of traditional agricultural labour relations, seasonal unfree labour and bonded labour systems were found to be continued in spite of new avenues for income and measures to abolish bonded labour system. The ritual significance of one form of seasonal unfree labour, (*thalavasage*) has declined. However, its practice in the present form was seen to be extending beyond the village for secular reasons.

With regard to bonded labour system it was seen that the coming of a cash crop itself, in more than one way, had acted as a source for its continuity. An increased demand for labour, either permanent or casual wage labour round the year, had encouraged the landowners to offer attractive terms and conditions to lure labourers into bonded service. An opportunity to sharecrop in sericulture was an incentive for a household to attach one of its male members as bonded labourer. Such incentives may be viewed as changes in the bonded labour system. Thus, farmers were now willing to pay more advance money as wages to the labourers, liberalise the working conditions, and more importantly, provide an opportunity for the attached labourer to gain his independence at the end of the fixed term. The last mentioned came about as a result of the disappearance of *baddi* form of bonded labour system (bonded against interest on a loan). Instead, the labourer was also given a small plot of land for cultivation of food crops or an opportunity to sharecrop. Thus, apart from the changes in the land economy (land reforms, sericulture, increased demand for labour, etc.,) ineffective credit policies and legal measures have also aided the continuity of these two institutions involving varying degrees of unfree labour.

The increased importance of exchange labour system can be viewed as an attempt by the farmers to organize themselves to overcome short supply of labour. This also signifies a continued interdependence of different sections of rural population in the process of cultivation of land. An increased commercial orientation as a result of sericulture has not affected any significantly this aspect of agrarian relations. Such exchange relationships, moreover, is found not only in the economic sphere, but also in all social and ritual spheres of lives in the village. The case of part-time sericulturists needs to be specifically recalled here. Being landless or not having land suitable to raise mulberry crops, they rear silk worms and harvest a crop or two of cocoons entirely with the help of other registered sericulturists. Such a part-time sericulturist hires all appliances from others, and pays for it and to the labourers hired after he receives money from the cocoon market. Thus, for some, apart from borrowing money, part-time sericulture offers a good chance of raising money to meet a contingency.

The picture that emerges from the analysis so far is of a process of "change and continuity" as against mere change or continuity in agrarian relations. This leads me to the second question I posed my self at the beginning of this chapter: What bearing does this process has on our understanding of social change in rural India? From the survey of literature (see Chapter 1), I had identified two broad types of conclusions emerging from those studies regarding changes in agrarian relations. First of them consider agrarian relations to be 'changing' from a traditional, customary, and interdependent relations to a commercial, conflicting and impersonal relations. The leading factors associated with such a transformation were identified to be land reforms, new technology, commercialisation of agriculture, and legislation abolishing some of the traditional systems of labour. In short, it was argued by those who uphold this perspective that as a result of land reforms, new technology, legislations etc., the traditional character of agrarian relations had changed. The second type of conclusions, although not an exact opposite of the first, refers to large scale distortions and inadequate or faulty implementation of the policies. Such distortions have led to the persistence of the traditional forms of agrarian relations.

On the basis of my study in Rajapura, I suggest a third perspective — "change and continuity" in agrarian relations as being a single process, and as viewed over a period of time. In other words, changes from certain forms of traditional agrarian relations to the commercial (or "modern") type of agrarian relations have occurred, but at the same time, many other forms of traditional agrarian relations are found to continue.

Studies on rural India have generally tended to pursue an unilinear approach in analyzing the process of social change. They tend to assume that either a given set of socio-economic factors produce a "total" change in a society, or have been change-resistant. In their eagerness to study and demonstrate changes in rural India scholars have rarely attempted to explain why some elements have persisted while the others have changed. In this sense, this study is unconventional in that it not only tries to locate the forces of change, but also identifies the patterns of continuity.

Notes

1. With the changes in the political leadership and in the government in the state, he lost his position since he was a camp follower of another leader.
2. When making this observation I am aware that applications for confirmation of occupancy rights refer to a given survey number of the land although the same survey number may have been sub-divided after each partition through generations, and further, the same survey numbered land may have been parcelled in smaller plots, and different tenants may have been cultivating them.
3. For a detailed account of *jajmani* relations, see Karanth (1987a), and on credit relations see (1988, and forthcoming).
4. Most accounts of *jajmani* relations have been silent on this aspect of the system. In Rajapura it was found that payment of wages annually was in relation to a scale of land owned by the patrons (eight seers of ragi per acre of land owned, and two bundles of straw) towards blacksmith, priest musician, etc., or in relation to the number of adult couples in a household (eight seers of ragi and two bundles of straw per every adult married couple) towards washerman, barber, etc.

Bibliography

Aggarwal, P.C., 1971, Impact of Green Revolution on Landless Labourers: A Note. *Economic and Political Weekly*, VI (47):2363-5.

———, 1975, Some Observations on the Changing Agrarian Relations in Ludhiana, Punjab. In V.R.K. Paramahamsa (Ed.) *Changing Agrarian Relations in India*. Hyderabad : National Institute of Community Development.

Alexander, K.C., 1975, Changing Labour-Cultivator Relations in South India. In V.R.K. Paramahamsa (Ed.) *Changing Agrarian Relations in India*. Hyderabad: National Institute of Community Development.

Aziz, Abdul and H.G. Hanumappa (Eds.), 1985, *Silk Industry: Problems and Prospects*. New Delhi: Ashish Publications.

Aziz, Abdul and Sudhir Krishna, n.d., *Land Reforms in Karnataka*. New Delhi: Sage Publications. (Forthcoming)

Bardhan, P and Ashok Rudra, 1980, Types of Labour Attachment in Agriculture: Results of a Survey in West Bengal 1979. *Economic and Political Weekly*, XV (35): 1477-84.

Beals, A.R., 1970, Namhalli 1953-1966: Urban Influence and Change in Southern Mysore. In K.Ishwaran (Ed.), *Change and Continuity in India's Villages*. New York: Columbia University Press.

Beteille, Andre, 1974, *Studies in Agrarian Social Structure*. New Delhi: Oxford University Press.

Breman, Jan 1974, *Patronage and Exploitation: Changing Agrarian Relations in South Gujarat, India*. Berkely: University of California Press.

Brown, R., 1970, *Seeds of Change: The Green Revolution and Development in the 1970s*. New York: Praeger Press.

Buchanan, F.H., 1807, *A Journey from Madras through the Countries of Mysore, Canara, and Malabar*. (Vols. 2 and 3). London: T. Cadell and W. Davies.

Byres, T.J., 1974, Land Reform, Industrialization and the Marketed

Surplus India: An Essay on the Power of Rural Bias. In David Lehman (Ed.). *Agrarian Reform and Agrarian Reformism.* London: Faber and Faber.

Charsley, Simon, 1980, A Silk Market in Karnataka. *Economic and Political Weekly*, XV: 1755-64.

———, 1982, *Culture and Sericulture: Social Anthropology and Development in a South Indian Livestock Industry.* London: Academic Press.

Chaudhari, D.P., 1974, New Technologies and Income Distribution in Agriculture. In David Lehman (Ed.), *Agrarian Reform and Agrarian Reformism.* London: Faber and Faber.

Chauhan, B., 1967, *A Rajasthan Village.* New Delhi: Associated Publishing House.

Correspondent, 1980, The Sham of Minimum Wages. *Economic and Political Weekly*, XV(24825): 1044-5.

Dalawai, Hussain, 1977, Maharastra: Bonded Labour Continues. *Economic and Political Weekly*, XII (22):868.

Damle, C., 1989, Impact of Tenancy Legislation and Changing Agrarian Relations: The Case of Dakshina Kannada District, Karnataka. *Social Scientist*, XVII(11-12):83-97.

Dasgupta, Biplab, 1977, India's Green Revolution. *Economic and Political Weekly*, Annual Number, XII(6,7 & 8): 241-60.

Department of Revenue, Govt. of Mysore, 1879, Sanctioning Proposals for Revisions of the Assessment of the unsettled Positions of the Magadi Taluk, Bangalore District. In *Mysore Settlement Reports*, Vol. II, Bangalore: Mysore Government Press.

Editor, 1978, Village Artisans: In Debt and Bondage. *Economic and Political Weekly*, XIII(36): 1007-8.

———, 1979, Deeper in Debt. *Economic and Political Weekly*, XIV(21): 893.

Eisenstadt, S.N., 1970, Prologue: Some Remarks on Patterns of Change in Traditional and Modern India. In K. Ishwaran (Ed), *Change and Continuity in India's Villages.* New York: Columbia University Press.

Epstein, T.S., 1962, *Economic Development and Social Change in South India*, Manchester: Manchester University Press.

———, 1967, Customary Systems of Rewards in Rural South India. In George Dalton (Ed.), *Economic Development and Social Change.* New York: The Natural History Press.

———, 1973, *South India: Yesterday, Today and Tomorrow.* London: The MacMillan Press Ltd.

Epstein,T.S. *et al*, 1983, *Basic Needs viewed from Above and Below: The Case of Karnataka State, India.* Paris: Organization for Economic Cooperation and Development (ECD).

Frankel, Francine, 1971, *India's Green Revolution: Economic Gains and Political Costs*. Princeton, N.J. : University Press.

Frykenberg, R.E. , 1977a, Introduction: Analytical and Historical Perspective. In R.E. Frykenberg, (Ed.), *Land Tenure and Peasant in South Asia*. New Delhi: Orient-Longman.

———, 1977b, The Silent Settlement in South India. In R.E. Frykenberg, (Ed.), *Land Tenure and Peasant in South Asia*. New Delhi: Orient-Longman.

George, P.T., 1970, Land System and Laws in Mysore State. *Artha Vijnana*, XII (1 & 2).

Gopal, S., 1971, *Tipu Sultan's Mysore: An Economic Study*. Bombay: Popular Prakashan.

Gould, H.A., 1970, Is the Tradition and Modernity All Bad?. *Economic and Political Weekly*, Special Number, V (29, 30 & 31): 1171-6.

Govt. of Mysore, 1905, *List of Villages in the Mysore State*. Bangalore: Government Press.

Guerfield, Joseph R., 1967, Tradition and Modernity: Misplaced Polarities in the Study of Social Change. *American Journal of Sociology*, 72 (4): 351-62.

Hanumappa, H.G.(Ed.), 1993, *Society, Economy and Sericulture*. Bombay: Himalaya Publications.

Harper, Edward B., 1959, Two Systems of Economic Exchange in Village India. *American Anthropologist*, 61(5): 760-78.

Harris, John, 1980, Why Poor People stay Poor in a Rural South India?. *Development and Change*. XI:33-64.

Henchette, S., 1972, Festivals and Social Relations in Mysore Village. *Economic and Political Weekly*, Special Number, VII (31, 32 & 33): 1517-22.

Herring, Ronald J., 1977, Land Tenure and Credit-Capital Tenures in Contemporary India. In R. Frykenberg (Ed.), *Land Tenure and Peasant in South Asia*, New Delhi: Orient-Longman.

———, 1980, Abolition of Landlordism in Kerala: A Redistribution of Privilege. *Economic and Political Weekly*, Review of Agriculture, XV(26): A59 - A69.

Hetne, Bjorn, 1978, *The Political Economy of Indirect Rule: Mysore 1991-1949*. New Delhi: Ambika Publications.

Hill, Poly, 1982, *Dry Grain Farming Families: Housaland (Nigeria) and Karnataka (India) Compared.* Cambridge: Cambridge University Press.

Jodha, N.S., 1971, Land based Credit Policies and Investment Prospects for Small Farmers. *Economic and Political Weekly*, VI (44): A143-A148.

Jose, A.V., 1978, Real Wages, Employment and Income of Agricultural Labourers. *Economic and Political Weekly*, XIII (12): A16-A20.

Joshi, P.C., 1975, *Land Reforms in India: Trends and Prospects.* New Delhi: Allied Publishers.

———, 1978, Land Reforms Implementation and The Role of Administrator. *Economic and Political Weekly*, XIII (39) : A-78 - A-83.

Karanth, G.K., 1981, *Rural Youth: A Sociological Study of a Karnataka Village.* New Delhi: Concept Publishing Company.

———, 1987a, New Technology and Traditional Rural Institutions: The Case of Jajmani Relations. *Economic and Political Weekly*, XXII (51):2217-24.

———, 1987b, *Primary Agricultural Credit Cooperatives in Karnataka: A Case Study.* Delhi: Department of Sociology, Delhi University (Mimeo).

———, 1988, *Formal Cooperatives and Informal Cooperation in Rural Karnataka: Lessons From the Informal to the Formal Cooperatives*, (Paper Presented in VIIth World Congress of Rural Sociology, Bologna, Italy. [June 28 - July 2, 1988].

———, n.d., Formal Cooperatives and Informal Cooperation in Rural Karnataka: A Micro Perspective. In B.S. Baviskar and D.W. Attwood (Eds.), *Cooperative Enterprises and Rural Development in India.* (Forthcoming)

Kessinger, Tom, 1974, *Vilayatpur 1848-1968.* Berkley: University of California Press.

Kulkarni, S.D., 1979, Bonded Labour and Illicit Money Lending in Maharashtra: Review of a Report. *Economic and Political Weekly*, XIV (X):561-4.

Kumar, Dharma, 1961, *Land and Caste in South India.* Cambridge: Cambridge University Press.

———, 1962, Caste and Landlessness in South India. *Comparative Studies in Society and History*, IV:337-63.

Ladejinsky, W., 1977, *Agrarian Reform as Unfinished Business: The*

Selected Papers of Wolf Ladejinsky. In L.J. Walinsky (Ed) New York: Oxford University Press.

Lal, Deepak, 1976, Agricultural Growth, Real Wages and Rural Poor in India. *Economnic and Political Weekly*, XI (26).

Lakshminarayana, H., 1977, Changing Conditions of Agricultural Labourers. *Economic and Political Weekly*, XII (43):1877-90.

Lockwood, David, 1964, Social Integration and System Integration. In G.K. Zellschana and W. Hirsch (Eds.), *Explorations in Social Change*. London: Routledge and Kegan Paul.

MacLachlan, M.D., 1983, *Why they did not Starve: Biocultural Adaption In a South Indian Village*. Philedelphia: Institute for the Study of Human Issues.

Mandelbaum, David, 1972, *Society in India*. Bombay: Popular Prakashan.

Manjappa, Kadidal, 1984, Land Reforms in Karnataka-Part I: The Origin of Legislation; Part II. Plea for Appelliate Tribunals: *Deccan Herald*, Bangalore, January 4 & 5.

Manor, James, 1980, Pragmatic Politics in Regional Politics: The Case of Devaraj Urs. *Economic and Political Weekly*, XV (5, 6, and 7):201-13.

Marx, Karl and F.Engels, 1976, *Selected Works*, Moscow: Progress Publishers.

Mencher, J.P., 1970, Change Agents and Villagers: Their Relationships and Role of Class Values. *Economic and Political Weekly*, V (29, 30 & 31): 1187-97.

———, 1978, *Agriculture and Social Structure in Tamil Nadu: Past Origins, Present Transformations and Future Prospects.* Delhi: Allied Publishers.

Michie, B.H., 1973, Variations in Economic Behaviour and the Green Revolution: An Anthropological Perspective. *Economic and Political Weekly*, VIII (26): A-67 - A-75.

———, 1978, Baniyas in the Indian. Agrarian Economy: A Case of Stagnant Enterpreneurship. *Journal of Asian Studies*, 37 (4): 637-52.

Ministry of Agriculture, 1976, Report of the National Commission on Agriculture, Part XV, Agrarian Relations. New Delhi: Government of India.

Ministry of Social Affairs, 1969, *The Causes and Nature of Agrarian Unrest*. New Delhi: Research and Policy Division, Government of India. (Mimeo).

Mukherjee, Radhakamal, 1933, *Land Problem of India*. London: Longman Green and Co.

Murthy,Sachidananda, 1984, Loan Mela: Playing With Money. *The Week*, II (15): 16-21.

Myrdal, Gunnar, 1977, *Asian Drama, An Enquiry into the Poverty of Nations*, (Abridged Edition). Penguin: Hammondsworth.

Nataraj, Lalitha, 1980, Karnataka: Politics of Land Reforms. *Economic and Political Weekly*, XV (26):1093.

Neale, W.C., 1962, The Indian Peasant, the State and Economic Development. *Land Economics*, XXXVIII (4): 281-91.

———, 1979, Land is to Rule. In R.E. Frykenberg (Ed.), *Land Control and Social Structure in Indian History*, (Enlarged Indian Edition). New Delhi: Manohar.

Oommen, T.K., 1967, Rural-Urban Continuum Re-examined in the Indian Context. *Sociologia Ruralis*, VIII (1): 30-48.

———, 1971a, Agrarian Tension in India: The Case of Alleppey District. *Indian Journal of Industrial Relations*, VI (2).

———, 1971b, Green Revolution and Agrarian Conflict. *Economic and Political Weekly*, VI (26).

———, 1972, *Charisma, Stability and Change: An Analysis of Bhoodan and Gramdan Movement in India*. New Delhi: MacMillan Press.

———, 1974, *Peasant Organizations in India: The Case of Alleppey (Kerala)*. F.A.O. Survey Report, New Delhi. (Mimeo).

———, 1975, Agrarian Legislation and Movements as Sources of Social Change. *Economic and Political Weekly*, X (40).

———, 1977, Analysing Rural Social Change: A Perspective. In M.N. Srinivas *et al* (Eds.), *Dimensions of Social Change in India*. Bombay: Allied Publishers.

———, 1984, *Social Transformation in Rural India: Mobilization and State Intervention*. New Delhi: Vikas Publishing House Ltd.

Pani, Narendar, 1983, *Reforms to Pre-empt Changes: Land Legislations In Karnataka*. New Delhi: Concept Publishing Company.

Planning Commission, 1973, *Report of the Task Force on Agrarian Relations*. New Delhi: Government of India. (Mimeo).

Rajan, M.A.S., 1979, *Land Reform Laws in Karnataka*, Bangalore: Government of Karnataka.

Ranganathan, N.D., n.d., *A Bird's Eye View of Land Reforms in Karnataka*. Bangalore: Government of Karnataka. (Mimeo).

Rao, C. Hayavadana, 1929, *Gazetteer of Mysore*, III & IV. Bangalore, Government Press.

Rao, M.S.A., 1979, Sociology in the 1980s. *Economic and Political Weekly*, XIV (44):1810-14.

Ray, Ratnalekha, 1979, *Change in Bengal Agrarian Society, 1760-1850*. New Delhi: Manohar.

Roth, Nans-Dietor, 1979, Moneylender's Management of Loan Agreement: Report of a Case Study in Dhanabad. *Economic and Political Weekly*, XIV (28): 1166-70.

Rothermond, Dietar, 1977, A Survey of Rural Migration and Land Reclamation in India, 1885. *Journal of Peasant Studies*, IV (3):230-42.

Rudolf, L. and S. Rudolf, 1967, *The Modernity of Tradition: Political Development in India*. Chicago: University of Chicago Press.

Rudra, A., 1975, Loans as a Part of Agrarian Relations; Some Results of a Preliminary Survey in West Bengal. *Economic and Political Weekly*, X (28): 1049-53.

Sen, B., 1970, Opportunities in the Green Revolution. *Economic and Political Weekly*, Review of Agriculture, V (13): A33 - A40.

Shah, A.M., 1974, Historical Sociology: A Trend Report. In *A Survey of Research in Sociology and Social Anthropology*, I: 432-59.

Shah, Ghanshyam, 1978, *Agricultural Labourers: Are they Bonded?* Surat: Centre for Social Studies. (Mimeo).

Singh, Dawat and V.K. Singh, 1975, Changing Agrarian Relations in Uttara Pradesh: Some Observations. In V.R.K. Paramahamsa (Ed.), *Changing Agrarian Relations in India*. Hyderabad: N.E.C.D.

Singh, V.B., 1964, Agrarian Relations in India. In Ignancy Sachs (Ed.), *Agriculture, Land Reforms and Economic Development*. Warsaw: PWN-Polish Scientific Publishers.

Sinha, Amin, 1978, Legal Loopholes to Landlord's Rescue *Economic and Political Weekly*, XIII (42): 1758-60.

Sivakumar, S.S., 1978, Aspects of Agrarian Economy in Tamil Nadu: A. Study of Two Villages, (Part I, II & III). *Economic and Political Weekly*, XII (18, 19 & 20).

Sonachalam, K.S., 1970, *Land Reforms in Tamil Nadu*: Delhi: Oxford University Press.

Srinivas, M.N., 1962, *Caste in Modern India and other Essays*. Bombay: Asia Publishing House.

———, 1976, *The Remembered Village*. Delhi: Oxford University Press.

Srinivas, M.N., 1979, Some Reflections on Rural Development. *Mainstream*, Republic Day Issue, January 26.

Srinivas, M.N. and A.M. Shah, 1960, Myth of Self Sufficiency of Indian Village. *Economic and Political Weekly*, XII (37).

Thakurata, Shankar Pada, 1976, Impact of Industrialization on Agratian Economic Strucutre: Some Preliminary Observations. *Journal of Indian Anthrological Society*, XXI: 273-82.

Thorner P Daniel, 1976, *Agrarian Prospect of India*. Bombay: Allied Publishers.

Thorner, Alice and Daniel Thorner, 1962, *Land and Labour in India*. Bombay: Asia Publishing House.

Whitcombe, E., 1980, Whatever Happened to the Zamindars?. In E.J. Hobsbalwn et al (Eds.), *Peasants in History*. Calcutta: Oxford University Press.

Wolf, Eric , 1966, *Peasants*. New Jersey: Prentice Hall, Inc.

Index